Rural Electrification

Rural Electrification

Optimizing Economics, Planning and Policy in an Era of Climate Change and Energy Transition

Edited by

Najib Altawell

Contribution Editors

James Milne
Patrice Seuwou
Luisa Sykes

ELSEVIER

ACADEMIC PRESS
An imprint of Elsevier

Academic Press is an imprint of Elsevier
125 London Wall, London EC2Y 5AS, United Kingdom
525 B Street, Suite 1650, San Diego, CA 92101, United States
50 Hampshire Street, 5th Floor, Cambridge, MA 02139, United States
The Boulevard, Langford Lane, Kidlington, Oxford OX5 1GB, United Kingdom

Notices
Knowledge and best practice in this field are constantly changing. As new research and experience broaden
our understanding, changes in research methods, professional practices, or medical treatment may become
necessary.

Practitioners and researchers must always rely on their own experience and knowledge in evaluating and
using any information, methods, compounds, or experiments described herein. In using such information
or methods they should be mindful of their own safety and the safety of others, including parties for whom
they have a professional responsibility.

To the fullest extent of the law, neither the Publisher nor the authors, contributors, or editors, assume any
liability for any injury and/or damage to persons or property as a matter of products liability, negligence or
otherwise, or from any use or operation of any methods, products, instructions, or ideas contained in the
material herein.

Library of Congress Cataloging-in-Publication Data
A catalog record for this book is available from the Library of Congress

British Library Cataloguing-in-Publication Data
A catalogue record for this book is available from the British Library

ISBN: 978-0-12-822403-8

For information on all Academic Press publications visit our
website at https://www.elsevier.com/books-and-journals

Publisher: Brian Romer
Acquisitions Editor: Graham Nisbet
Editorial Project Manager: Chris Hockaday
Production Project Manager: Kamesh Ramajogi
Cover Designer: Greg Harris

Typeset by TNQ Technologies

This book is dedicated to all those who are working in the field of energy and rural electrifications - with the aim of supporting and developing a healthy environment and a better world.

Contents

List of figures

List of tables

List of boxes

Foreword

At its heart, energy is all about people. Billions of the world's population are dependent on energy to go about their daily lives safely, comfortably, and productively.

For millions of those people, it is also a part of daily life to work in a job that helps provide that energy.

They are the people who keep the lights on, the gas flowing, the forecourts full of petrol, and—increasingly—who develop the energy technologies that will help us tackle climate change and bring energy to the billions around the world still living without meaningful access to energy.

And that is crucial. Climate change and opening up access to energy to the world's most vulnerable populations—in sub-Saharan Africa, in particular—must be tackled together.

But I am a firm optimist and believe that humankind has the capability to resolve them. We should take encouragement from two things.

First, the international resolve is there. We saw this in 2015, with the agreement of the UN's Sustainable Development Goals, including the commitment to ensure access to affordable, reliable, sustainable, and modern energy. And later that year, we all celebrated the Paris Agreement, an extraordinary milestone with 197 countries signing up to keep the increase in global average temperature to well within 2°C of pre-industrial levels.

Second, our ability to innovate and develop technology has led to steep changes in human development before, and it can do so again. Although the world is still heavily dependent on oil, gas, and coal, the advances, and cost reductions being made in new technologies—solar in particular—are already benefiting countries and populations at all stages in their socioeconomic development. Watch this space for much, much more.

These challenges are existential, both at an individual level and a planetary level. But, if we can equip professionals, academics, policymakers, and the public with the very best knowledge, information, and evidence about energy, then informed, transformative responses can be put in place.

Through collaboration, we can ensure that the very best of humankind's creativity and scientific endeavors are put to work to meet the grand energy challenge.

Louise Kingham OBE FEI
Chief Executive
The Energy Institute

Preface

The idea to compile a book that covers various aspects of energy came about at the beginning of 2014, i.e. as soon as I completed a manuscript that dealt with biomass energy. Prior to 2014, it took me many years searching and examining various types of books that focused on the field of energy alone. However, the search itself made me discover that many of these books I managed to access did not provide a wider and comprehensive energy frame of work covered in one book. Years later, the idea of a full book covering energy began to take shape in my mind and I noted down on a piece of paper various related materials, each day. Soon enough, a gradual process for the proposed book started to take shape, mostly at the end of each lecture, I presented that dealt with energy, directly or indirectly. By the beginning of 2016, there were enough written materials that gave me the idea to contact other members of staff at GSM London to find out if they were interested in contributing to a newly proposed book dealing with energy. The staff were enthusiastic and interested in the work, and more specifically when they discovered it was relevant to their own field of work.

By the end of 2017, the materials submitted by some of the contributors sparked an interest to add additional chapters to the proposed book. In having said that, challenges began to appear as the work began to take a certain direction, and some members of the staff who committed part of their time to contribute to the book, either disappeared quietly or sent a letter of apology that they were no longer able to contribute to the book. Those contributors who did not manage to stay with us had either been offered a new position at another institute or had simply moved abroad.

The above meant that the original team of contributors got smaller, which meant the book editor and whoever was left of the original team had to work harder by working on additional materials that were not allocated to them before. This also meant that the proposed book contents, as well as the title, agreed with the publisher, had to change from "The Energy Book" to the present title "Rural Electrification: Optimizing Economics, Planning and Policy in an Era of Climate Change and Energy Transition." The change of path meant that a new contributor needed to join our team for further and more specialized energy inputs.

Despite some of the major challenges the contributors' team faced for more than 2 years, the manuscript continued to grow and develop into thirteen chapters covering a wide range of energy fields and systems. This successful outcome I simply attributed to the team's desire and determination in completing this work and on time. I am pleased, therefore, to say that what had been envisaged in mind many years ago has now been developed to the outcome we all wanted, i.e., the book you now hold in your hands.

N. Altawell
Editor

Acknowledgements

I convey my gratitude and appreciation to everyone who has contributed, directly and indirectly, to the successful completion of this book. I would like to thank Mr James Milne, Mrs Luisa Sykes and Dr. Patrice Seuwou for their support and effort over a period of more than 2 years and for the chapter contributions they have provided to this book.

I would like to thank Louise Kingham, Chief Executive of the Energy Institute, for contributing the opening section for this book. Once again, I am grateful to James Milne, Luisa Sykes and Sagar Das for reviewing the manuscript.

Thanks to all the staff at Coventry University (Greenwich and Dagenham campuses) for their encouragement over the past few months during the final stages of compiling this book.

Abbreviations

ADB	Asian Development Bank
API	American Petroleum Institute
ARA	Amsterdam-Rotterdam-Antwerp area for pricing petroleum products
bbl	Barrel of oil (159 L or 42 US gallons)
bcm	Billion cubic meters
BIPV	Building-Integrated Photovoltaics
bpd	Barrels per day (of oil production)
Btu	British Thermal Unit
CDM	Clean Development Mechanism
CES	Cryogenic Energy Storage
cm	Centimeter
CNG	Compressed Natural Gas
CO_2	Carbon Dioxide
degC/°C	(temperature measured in) Degrees Celsius
degF	(temperature measured in) Degrees Fahrenheit
DG	Distributed Generation
DTI	Department of Trade and Industry
DWT	Deadweight Tonnes (a measure of how much a ship can carry)
EC	Energy Carriers
EE	Energy Efficiency
EEG	Extension of the Existing Grid
EES	Electrochemical Energy Storage
EIA	Energy Information Administration
ES	Energy Storage
ESS	Energy Storage Systems
EWP	Energy White Paper
FID	Final Investment Decision
FOB	Free On Board (title in cargo passes at point of transshipment)
GE	Green Energy
GHG	Greenhouse Gas
GIIGNL	The International Group of Liquefied Natural Gas Importers
GSE	Government-Sponsored Enterprise
GTL	Gas-To-Liquid (manufacturing process)
HESS	Hydrogen-based Energy Storage Systems
ICE	Intercontinental Exchange (trading house in London, UK)
IEA	International Energy Agency
IOC	International Oil Company
IRENA	The International Renewable Energy Agency
km	Kilometer
kph	Kilometers per hour
LNG	Liquefied Natural Gas
m^3	Cubic meters

MES	Mechanical Energy Storage
mg/kg	milligrams per kilogram
MHPS	Micro Hydro Power System
mm	Millimeter
MMBtu	Million British Thermal Units
MMBtu	Millions of British Thermal Units
Mw	Megawatt
NBP	National Balancing Point (virtual trading point for UK natural gas)
NOC	National Oil Company
NYMEX	New York Mercantile Exchange
OECD	Organization for Economic Cooperation and Development
Ofgem	Office of Gas and Electricity Markets
OPEC	Organization of Petroleum Exporting Countries
OTC	Over-the-Counter (oil and gas trade deals not conducted via an exchange house)
ppm	Parts per million
PSA/PSC	Production Sharing Agreement or Contract
PV	Photovoltaic
R/T	Royalty/Taxation system
RE	Renewable Energy
REA2	Renewable Energy Analyzer Two
RES	Renewable Energy Sources
RET	Renewable Energy Technologies
ROC	Renewables Obligation Certificate
SBSP	Space-Based Solar Power
SD	Sustainable Development
SEB	State Electricity Board
SHP	Small Hydropower
SHS	Solar Home System
SMES	Superconducting Magnetic Energy Storage
SWER	Single-Wire Earth Return
tcm	Trillion cubic meters
TES	Thermal Energy Storage
toe	Tonnes of oil equivalent
TWh	Terawatt hour
ULCC	Ultra Large Crude Carrier
UNIDO	United Nation Industrial Development Organization
VLCC	Very Large Crude Carrier
WTI	West Texas Intermediate (benchmark crude oil for pricing US imports)

Energy's general introduction (fossil fuels & renewable energy)

Everything is energy, and that's all there is to it.

Albert Einstein.

The word "energy" is defined by the Oxford English Dictionary as "The strength and vitality required for sustained physical or mental activity," which simply means anything connected with activity, and without it, nothing can exist. Therefore, energy is whatever exists, the visible, and nonvisible in our world, and beyond, all is a form of energy; a solid matter is a form of energy at a different level of activity.

The word *energy* originates from the Greek language *Energeia*, meaning *activity*. However, while humans have managed to control and manipulate various forms of energy, their outputs in the usage of energy and applications have varied according to their standard of living throughout history. More than ever before, today's level of a nation's advancement and the quality of their living is measured directly based on the amount and type of energy they use (Figs. 1.1 and 1.2).

This simply means poorer countries need more energy to combat poverty, health issues, and education as this is needed at various stages of any nation, regardless of the conditions of the country.

Therefore, civilizations, according to the Kardashev scale, are measured based on the amount of energy that humans are able to use. According to Kardashev, there are three types of civilizations: type 1 civilizations can use and store all of the energy available on their planet, while type 2 civilizations can harness the total energy of their planet's parent star and the third type of civilization can control energy on the scale of its entire own galaxy (Kardashev, 1984). At the present time, humanity has not yet reached any of the above three level types. Possibly, within the next 100 years or so, we may reach type 1. What this means is that advancement in every aspect of our lives cannot be achieved without further development within the field of energy (Box 1.1).

At the same time, although the energy in peacetime and wartime provided advancement and destruction, the advancement factor was much bigger than the destruction factor; otherwise, we would not be in existence now. Even the destruction that humans caused by the usage of energy gave birth to innovations and developments in various fields. Therefore, the future positive and negative outcomes of energy

Rural Electrification. https://doi.org/10.1016/B978-0-12-822403-8.00001-1

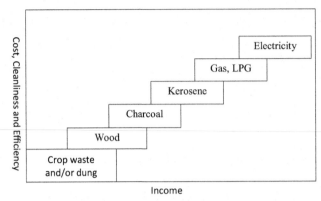

FIGURE 1.1 Standard of living illustrated in the form of "Energy Ladder" for rural India. *Redrawn and edited from the source Duflo et al., 2008.*

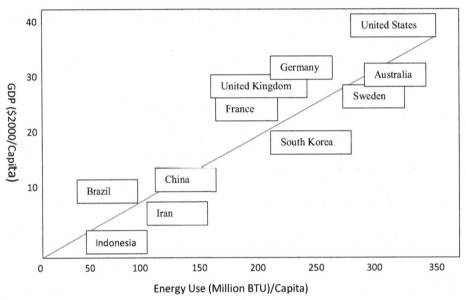

FIGURE 1.2 Standard of living illustrated in the form of energy usage. *Redrawn and edited from the source CIA, The World Factbook: BP plc.*

BOX 1.1 What is the difference between force and energy?

- Energy is an ability to operate or activate things while force is a method of transferring energy
- Energy and mass of a closed system is conserved, but there is no such conservation for force
- Force is a vector quantity while energy is a scalar

From "DifferenceBetween.Com". Difference Between Energy and Force (2011). https://www.differencebetween.com/difference-between-energy-and-vs-force/ (accessed 18 February 2020).

applications, as science and technology advance progressively, will certainly provide various possibilities to humanity in the future. Of course, future possibilities in this field are in our hands. That is why decision-makers, i.e., those who are in control of regulations and laws, as well as those who design and implement important applications of energy, play a major and vital role and will be responsible for the future outcomes for all of us.

One of the most important issues facing humanity today is the negative impacts of using a common form and a well-established source of energy: fossil fuels. The outcomes for burning fossil fuels has resulted in what has been termed as "Climate Change and Global Warming." What the above means is that what nature did over a period measured in geological time scales in reducing CO_2 from the atmosphere when the concentration of CO_2 during the above period was around 95% to 99%, we are now reversing the above process by releasing back the CO_2. This kind of emission happens via the process of burning fossil fuels. Microorganisms such as plankton helped to reduce CO_2 from the Earth's atmosphere long before humans existed on this planet.

One of the most important factors that contributed to increasing emissions of CO_2 was a certain type of "invention." What is the invention that may have accelerated global warming and climate change? you may ask.

Let us remember that the CO_2 concentration in the atmosphere before the industrial revolution was approximately 40% to 45% lower than it is today. Obviously, it was the invention of the combustion engine that was the starting point that led us to the present issue of climate change and global warming.

You may ask, "if we continue using fossil fuels, what are the consequences for the planet Earth"? The sharp rising of CO_2 in the atmosphere, together with other greenhouse gases, will form a thick blanket around the Earth, trapping heat further as well as changing the pattern of the climate drastically. Releasing greenhouse gasses continuously may change the Earth's climate, possibly leading it to become similar to the climate on the planet Venus where the concentration of CO_2 now is similar to the concentration of CO_2 on Earth in past geological ages. The average temperature on the planet, Venus, is approximately 465°C.

What are the major sectors around the world emitting a high percentage of CO_2 via man-made activities? The main sources of man-made CO_2 are from electricity and heat production, transport, the manufacturing and construction industries, refineries, and other sectors. Eighty percent of our energy is still sourced from fossil fuels, and rural areas, especially in developing countries, where the usage of fossil fuels is even much higher. Therefore, the focus on fossil fuels, renewable energy, and aspects connected to rural electrification is an important subject to shed light on, from both environmental issues and to the people in the countryside (Box 1.2).

BOX 1.2 The biggest energy challenges facing humanity

With the global population swelling and industrialization on the rise in developing nations, humanity's hunger for energy has reached unprecedented levels. More than half of our energy comes from fossil fuels extracted from deep within the Earth's crust. It is estimated that since commercial oil drilling began in the 1850s, we have sucked up more than 135 billion tonnes of crude oil to drive our cars, fuel our power stations, and heat our homes. That figure increases every day. But our gas-guzzling over the past two centuries has taken a potentially devastating toll on the planet. Burning coal, oil, and gas has been inextricably linked to the rising levels of greenhouse gases in the Earth's atmosphere and is a leading contributor to climate change. The world's scientists agree that we are on a path toward disaster that can only be stopped by weaning ourselves off our fossil fuel habit. But that leaves us with a problem. How do we ensure the lights stay on?

From George R. Harrison (1955). The Control of Energy. The Atlantic. https://www.theatlantic.com/magazine/archive/1955/09/the-control-of-energy/305024/ (accessed 18 February 2020).

1.1 Brief main aspects of renewable energy

The following is a summary of the various characteristics of RE. Full details of RE sources and systems can be found in Chapter 7.

1.1.1 Biomass

No Limitations.
Three Different Forms of energy sources.
Energy from Landfill Gas.
The Possibility Of Zero GHG Emission.
Reduction of CO_2 in the atmosphere.
Ash as a fertilizer and other purposes, e.g., for the cement industry and road building materials.
Hot-water boiler issues.
The Need for Land and Water.
Direct energy—Thermal, Mechanical, and Electrical—competitive.
Liquid Fuel for Combustion Engines.

1.1.2 Hydro

A constant supply of electricity at a very low cost.

Electricity can be supplied from SHP for more than one mile. No need for reservoir access to electricity that can be sold to the main power station. Environmental impact is the low cost of MHPS compared to the value of energy provided on a long-term basis. Summer time may have an impact on the amount of power produced. The amount of water flow may limit future power expansion. The source can be limited as it depends solely on water elevation, flood control and irrigation positive aspects on "aquatic" ecosystems, development around the local area siltation, soil erosion, soil and water salinity, obstruction of the free passage between oceans and rivers, weed growth, floods due to dam failures, spread of disease by organisms (stagnant water), and damage to natural resources (fish).

1.1.3   Ocean thermals

An abundant and clean source of energy.
The environmental impact is minimal.
Pumping deep seawater will require additional energy, such as the use of fossil fuel/diesel generators.
Produce fresh and cold water.
Expensive in terms of construction and maintenance.
Systems, less than 50 MW output, are not considered as viable.
Environmental Issues (concerns).
It can be used to help in the refrigeration process.
Air-conditioning system for domestic and commercial use.
Production of methanol, ammonia, and hydrogen.
OTEC-system built on the land is less costly than a system placed in the ocean.

1.1.4   Geothermal

Using the Earth's natural heat underground as a source of energy steam to drive turbines and for heating and cooling systems. Deep drilling—can be an expensive source of energy continuous and constant supply of energy—but not always!

1.1.5   Solar

Based on semiconductor solar panels, sunlight and/or sun heat for the production of electricity and/or heating/cooling require storage banks (large batteries) and/or back-up system. Weather conditions and location play a major role. Mostly suitable for small domestic applications. The term "low emission" has been given to this type of technology. The installation is simple. The maintenance cost is low (without the inclusion of batteries). It can operate successfully for many years without any intervention. It can function in harsh conditions (reliability). Large areas of land are needed to capture the sunlight. Charge batteries as a way of storing energy (night/not enough sunlight). No noise or by-products.

1.1.6   Wind turbine

Micro wind-turbine for domestic electricity supply. One of the fastest developing technologies worldwide, combined with an energy storage facility, it can provide power on 24 h/day basis. The total cost can be cheaper than a solar farm but more expensive than hydroelectricity production; depends on wind speed, location, season, and air temperature. Average capacity factor can be close or higher than 30% zero greenhouse gas emissions (not counting the emission produced via the manufacturing process of the device). High percentage of the hardware cost (for large wind turbines) is mostly spent on the tower designed to support the turbine. A wind turbine is suitable to be installed in remote rural areas. Wind turbines can also be used for water pumping. Noise, shadows, and light fluctuations can be a negative aspect if wind turbines are constructed close to residential areas.

1.2 Hybrid systems

Applying the strong points of multiple technologies.
Hybrid systems should be close to optimal in both performance and cost.
The capital cost is high (e.g., cost of solar, cost of batteries, and cost of maintenance).
Better reliability of power supply.
Need technical and engineering know-how.
Ideal as a back-up power generating system.

1.3 Sustainability

The "World Commission on Environment and Development," 1987 defined sustainability as:

> *"Meeting the needs of the present without compromising the ability of the future generations to meet their needs".*

The above definition in its generic content has been adapted in specific terms in relation to the oil and gas three stages, i.e., exploration, production, and distribution. Although sustainable development is often talked about in various parts of the world, including by oil and gas companies, the sad truth is that very little has been achieved in this field, and even some of the work completed in this direction is, in many cases, ineffective and/or its implementation improper. In some cases, the above poor result is simply due to a lack of funding (UNCTAD, 2017). Whether it is a lack of funding or neglect, the environment and living species (including humans) were and are the main victims of what is happening, especially what oil and gas companies have done so far, for more than 100 years of unsustainable approaches. Therefore, further action on the ground is needed on how to implement and develop sustainable development anywhere in the world, i.e., by simply reducing the negative impacts and investing heavily in sustainable production, or even better, in sustainable energy that involves the local community as an important part of this sustainable development.

Oil and gas companies invest in sustainable energy (e.g., renewable energy) for the following main reason: as they are in the energy business, they know that fossil fuels (e.g., crude oil) will run out in the future. There are grants and tax reductions available from various governments around the world (especially in the West) for any business, including oil and gas companies that can benefit financially from these government schemes. Oil and gas companies can claim that they are looking after the environment by investing in sustainable energy, which can be used as part of their marketing publicity as they deal with different governments around the world. The above can be summarized in the following points:

1. Pressure from stakeholders, shareholders, and governments
2. Climate change regulations at the country level and International Climate Change agreements
3. Establish eco credentials and enhanced reputation
4. Diverse energy investments ahead of estimated peak demand
5. Availability of grants and tax reduction from various governments around the world to invest in clean energy and pollution abatement technology
6. Marketing and PR campaigns to promote brand awareness of RE companies

Consumption of oil will continue to rise, matching the increase of the world population and the need for a better standard of living, especially in the poorer developing countries. Sustainable sources of energy need further investments and supporting regulations to speed-up implementation, in particular when it comes to the distribution network and affordable process. Other factors can be listed in the following points:

• Implementing energy-efficient processes
• Reducing flaring of gases
• Reusing the captured gas
• Investing in research and development
• Developing core competencies in sustainable energy development
• Expanding to new markets with sustainable energy products (Box 1.3).

1.4 Oil consumption

To provide a picture of the present situation related to oil consumption, a focus on three major consumers of oil will provide some of the important issues and factors impacting on our world. The three countries are the USA, China, and India.

BOX 1.3 The control of energy

One pound of uranium carries more releasable energy than 1500 tons of coal, and the solar energy that reaches the earth in a single day is equivalent to that released by two million Hiroshima A-bombs. Better control of these and other forms of energy is basic to man's progress.

From George R. Harrison (1955). The Control of Energy. The Atlantic. https://www.theatlantic.com/magazine/ archive/1955/09/the-control-of-energy/305024/ (accessed 18/02/2020).

1.4.1   USA

The USA has become an exporter (as well as a net importer) of crude and liquefied natural gas since the ban on the export of crude oil was lifted during 2016; the USA is also one of the leading countries in the world as a consumer of fossil fuels (IEA, 2017a,b). According to the EIA (2010), the USA consumes and will continue to consume for the foreseeable future, more energy from fossil fuels than from any other energy source. For example, the USA petrochemical industry depends heavily on oil derivatives for the production of a multitude of essential products. In addition, shale technologies have released previously inaccessible oil and gas reserves, heralding a new production bonanza for the US oil and gas companies. In spite of persistently low prices in the period between 2014 and 2017, the continuous rise in US oil and gas production was possible as a result of cost reduction connected to technological innovation. However, comments made by some experts, such as Fatih Birol, IEA Executive Director, during an interview on 16th March 2018 by CNCB TV Channel reflect a different view:

> *"Upstream investment shows little sign of recovering from its plunge in 2015–2016, which raises concerns about whether adequate supply will be available to offset natural field declines and meet robust demand growth after 2020".*
>
> *(WEO-IAE, 2017c)*

On the other hand, with developments in the production of tight oil, the IEA predicts a better future:

> *"A remarkable ability to unlock new resources cost-effectively pushes combined United States oil and gas output to a level 50% higher than any other country has ever managed; already a net exporter of gas, the US becomes a net exporter of oil in the late 2020s".*
>
> *(WEO - IEA, 2017a,b)*

1.4.2   China

Although China has been a net importer oil and gas since1990, it has already surpassed the USA when it comes to the consumption of fossil fuels, particularly oil (EIA, 2015). This is mainly due to the unprecedented pace of economic development in China, which has increased the demand for energy drastically within the past 20 years. For sustaining rapid economic growth, energy has become one of the main tipping points. It means that continuous and increasing demand for oil from China has imposed immense pressure on energy resources creating risks to global energy security if no genuine international cooperation within the energy markets, in general, and oil in particular, actively takes place. The above is one of the reasons why some experts have frequently connected high oil prices experienced in the past with increasing demand from China. An IEA prediction during 2007 estimated that the world energy needs will be more than 50% higher in 2030 than today's needs—if global governments continued with their present policies—as

China and India together account for 45% for this increase alone (IEA, 2007). However, in today's demand profile, the prediction was not completely accurate. Demand for crude oil, in particular, was not stimulated by low oil prices as had been anticipated, but poorer nations may have benefited, to a certain extent, from the market oversupply of oil, i.e., reduction in prices per barrel. Fast-forwarding to 2018, the increase in oil prices may indicate the beginning of the recovery for the oil market, although experts considered this kind of price adjustment/barrel to be rather an artificial outcome than a reflection of actual market stability. Therefore, the future demand for oil is expected to increase further, despite the commercial establishment of other nonfossil fuel sources of energy. As a result of the above higher oil demand, i.e., a greater dependency on oil producer countries, such as the unstable region of the Middle East, may increase competition among oil importers for a higher share of crude oil from these countries, as well for the newly discovered oil reserves in Africa and South America. Oil resources for China have become a major factor for its present and future development; without constant higher energy input, the fast pace of industrialization will come to a halt, sooner rather than later. By examining China's energy resources, regulations, policy, and the actual needs, as well as international concerns resulting from growth in energy demand, a clearer picture will emerge as to what kind of actions should be considered by the international community to resolve possible future energy challenges and market destabilization.

On the other hand, reducing costs, and encouraging the development of energy sources other than fossil fuels, financial support, and relevant institutional, governmental policies are vital for the initiation and construction of all kinds of sustainable energy projects. Whether for electrification programs and/or transportation purposes, the effort should be based purely on the economic development of the local population, i.e., a transparent agenda directed toward an actual benefit for urban and rural communities. Technical issues, whether in the form of design and/or standard requirements, should be developed and adapted to fit with the local and national needs. Also, minimizing electricity losses during distribution, in the form of load and load mix, is another way of reducing cost and ensuring power supply. Subsidies and tariff structures, whenever these are applicable, are also other essential criteria, which will help in the development of China's electrification program.

1.4.3   India

India's oil consumption, presently a large part of its growing energy requirement, is imported from other parts of the world. This is despite the government's heavy investment in oil and gas exploration to help reduce the dependency on foreign energy sources. As an example, India's estimated annual GDP growth rate of more than 8% from 2005 to 2006serves as a good indicator of the level of development that took place during the above period (Sengupta). At the present time, however, India is slightly below previous predictions, although catching up very fast with a GDP forecast of 7.5% for both 2019 and 2020. (World Bank, 2018).

The potential in development and higher energy demand, therefore, may continue to rise for the foreseeable future, depending on India's economic stability and international politics. This kind of ascending demand for energy is a natural occurrence in order to fulfill the increasing scale of development, and consequently, for the provision of a higher standard of living, i.e., energy supply in various forms will always be needed. India's fast development and its lagging behind in rural electrification aspects make the urgency to balance the general growth with the need for distribution of energy, i.e., electrification, more urgent and vital than ever before.

Finally, the Indian government has introduced a number of incentives, mainly for the purpose of attracting investment in the field of sustainable energy. These incentives range from grants, subsidies, policy support, frameworks for regulatory and legislative aspects, consultations and sources of finance, research and development, planning and resource assessments, and help in upgrading existing energy generating technologies.

1.5 What are the impacts of global warming and climate change?

✔ Frequent heatwaves and water shortages
✔ Sea level rises:
✔ Population displacement
✔ Property losses more frequent/on a larger scale:
✔ Damage caused by storms, floods, and wildfires
✔ Farms, forests, and fisheries will be less productive
✔ Species extinction
✔ Spread of infectious diseases
✔ Heart and respiratory problems (due to higher concentrations of ground-level ozone)

With the present vast volume of fossil fuels production around the world, it is common knowledge that fossil fuels will run out, possibly within the next 70–80 years from now. The man who predicted in 1949 that the crude oil would be short-lived is M. King Hubbert. In 1956, he predicted that peak oil production in the USA would occur during the year 1970 (Hubbert's Peak). His prediction became fact as the USA, instead of being the world's first oil exporter prior to the decade of 1970, became an oil importer. However, horizontal drilling and fracking technologies over the past few years may have helped in reducing imports and have even turned the USA into a net exporter of oil and gas. This new technology may have revived oil production in the USA, but this is a short-lived situation, as this new method of oil production will eventually exhaust what is available at the present time in the USA, approximately within the next 10 years.

An important question can be raised concerning natural energy resources: This question is relevant to producers of fossil fuels in underdeveloped countries. Some of these countries are economically poor, although they have rich resources, such as crude oil.

"One of the surprising features of modern economic growth is that economies abundant in natural resources have tended to grow slower than economies without substantial natural resources".

(Sachs and Warner, 1997)

For some countries (rich in oil and gas), research shows that the boom has produced a negative outcome in the country rather than the opposite, e.g., the Democratic Republic of the Congo, Equatorial Guinea, Nigeria, and Niger. The main symptoms involve "the Dutch disease," rentier-state dynamics, rent-seeking behavior, armed violence, and corruption. If governments do not implement policies beyond those already planned between now and 2030, it is projected that:

- Energy consumption will increase by over half (53%)
- The energy mix will remain fairly stable and dominated by fossil fuels (80% share)
- Energy-related CO_2 emissions will increase by over half (55%), and large populations of the world's poor will continue to lack access to electricity (about 1.5 billion) and modern cooking and heating services (about 2.5 billion). Over 70% of this growth is expected to come from developing countries

New policies are needed to further reduce the emission of CO_2 in particular within the developing countries, as their populations are increasing rapidly, and as a result, a higher percentage of energy will be needed for their health and economic development, as well as to combat poverty.

Developing Countries: Growth in energy consumption are expected to be power generation (35%), industry (15%), transport (12%), and buildings (6%), in developing countries, OECD Countries: power generation (11%) and transport (6%).

To combat some of the challenges related to economic development in underdeveloped countries and to minimize the impacts on the environment, the following should be examined:

- Corruption issues in developing countries
- Further awareness and education of local populations
- Investments in renewable sources of energy
- Implementing energy-efficient processes
- Reducing flaring of gases
- Reusing the captured greenhouse gases
- Investing in research and development
- Developing core competencies in sustainable energy development
- Expanding to new markets with sustainable energy products

- Energy efficiency (reduction in wasted energy) - the commercial application of clean energy in developing countries is vital in reducing GHG emissions.
- Improving efficiency - reducing CO_2 emissions should be the first approach in these high growth areas; cost-effective at the time of new construction rather than at later retrofit stages.

1.6 Energy storage systems (The sixth dimension)

Why is there the need for ESS (Energy Storage Systems)? All living species, including us, have energy storage systems. Without energy storage systems, life cannot exist. The same thing is true about a regular supply of energy, whether in the form of electricity or general power supply, energy storage systems in both cases are vital. There are some old and new storage technologies that will be discussed in the following pages, as well as ideas and new approaches. Benefits may include: Cost reduction, having a back-up, and access to electricity in the form of energy, which can be stored, efficiency and reliability is an important factor when it comes to the smart grid.

1.6.1   ESS examples

1. Batteries: Solid-State Batteries - a range of electrochemical storage solutions, including advanced chemistry batteries and capacitors. Flow Batteries - batteries where the energy is stored directly in the electrolyte solution for longer cycle life and quick response times.
2. Compressed air: Ambient air is compressed and stored under pressure in an underground cavern.
3. Flywheel: This can be described as a mechanical battery, in that it does not create electricity, it simply converts and stores the energy as kinetic energy until it is needed.
4. Hydrogen fuel cells: Electricity can be converted into hydrogen by electrolysis. The hydrogen can then be stored and eventually re-electrified.
5. Pumped water: Pumped hydroelectric storage facilities store energy in the form of water in an upper reservoir, pumped from another reservoir at a lower elevation.
6. Electric Vehicles: As around 95% of cars are parked, especially during night time, batteries in electric vehicles could be used to let electricity flow from the car to the electric distribution network and back.
7. Molten Salt: Heated molten salt (by the sun) heats a tank of water in order to generate hot steam to be used to turn turbines to generate electricity.

Other technologies, including possible future approaches, may include energy stored by bending/deforming, i.e., the energy released when the material returns to its original shape; nanotube springs made from carbon nanotubes, and wires made from copper wire surrounded by two super-capacitor layers created by special nano-whiskers storing energy within the internal space of matter; recycling energy output from lights, lifts, and movement in general; using seawater movement that results from the gravitational force of the moon; finally, designing an efficient, commercially viable photo-synthesis machine.

Additional details related to ESS can be found in Chapter 12.

1.7 Type of sustainable energy needed

Coal, oil, and natural gas are all forms of fossil fuels, without which large parts of our technology industries could not exist. Can we use nuclear power instead of fossil fuels in order to reduce CO_2 emissions? Is nuclear power the missing energy we need? After all, there is no greenhouse gas emission when generating electricity from nuclear power? The challenges of the radioactive waste produced, the high cost involved in building a nuclear power station, security issues in case dangerous materials fall into the wrong hands (dirty bombs) and maintenance costs are all important factors that lead us to the conclusion that generating electricity from nuclear power on a large scale may not be the answer, at least within the near future.

What about renewable energy? The free natural resources for this kind of energy are available all around the world, so why do not power generating companies or governments decide to use these sources instead of using fossil fuels? Obviously, there were and still are many factors and challenges that may not have existed at the time of inventing the combustion engine. Past data shows us that the price of oil was low and environmental issues were never considered, and neither was serious consideration given to the fact that fossil fuels, in general, and oil in particular, are not sustainable, i.e., what fossil fuels we have inside the Earth's reserves will eventually run out. So how did the idea of renewable energy come about, and when?

During the war between Israel and Egypt in 1973, King Faisal of Saudi Arabia at that time decided to support Egypt by shutting-off the oil supply to the West. The above action shocked the oil market, and the price per barrel jumped from $2 to $25 following the king's decision. While the US government never expected that Saudi Arabia would consider stopping the oil supply, their response came by trying to implement a new approach with a new policy related to energy supply in general and crude oil in particular. Regarding energy supply, the idea that the USA should generate its own energy was an important decision by its government. This was followed by the conclusion that the USA's crude oil supply from the Middle East may not be reliable, and therefore, a new home-grown source of energy was considered: this was when the term "biomass" as a source of energy was coined. The USA is one of the top producers of corn, and starch from corn can be converted to ethanol, similar to Brazil's production of the above fuels from sugar cane.

However, the above approach in producing fuels from corn failed badly, mainly due to the higher cost of corn as it skyrocketed when additional demand on corn to convert to fuel led to companies deciding to produce this type of fuel, in addition to the demand for corn as a source of food and as a feed for animals in the USA and outside it.

The other action that the US government has taken to secure energy supply is stockpiling oil reserves on a much larger scale than ever before in the history of the USA.

The renewable energy approach, therefore, began during the early part of the 1970s, although renewable energy, such as hydropower, had been known and implemented in the past for many years long before the above date. The principles of some of these types of renewable energy can be traced back thousands of years. However, despite the fact that a number of these renewable energy systems were recently established as a commercial source of energy, the struggle for renewable energy systems to establish itself on the commercial market has been slow and difficult. This has been mainly due to the strong competition from fossil fuels. To illustrate as to why this kind of new system, i.e., renewable energy systems, did not manage to establish themselves faster than the other new systems can be examined in the following example:

As mentioned previously, during the early 1970s, the first commercial approach for a renewable energy system began. Midway through the 1980s, the first commercial mobile phone appeared on the market. Approximately 15 to 16 years later, renewable energy systems, such as energy crops, commenced. Within only a few short years, mobile phones managed to establish themselves across the globe. Whether you live in the West or various other parts of the world, almost everyone now has a mobile phone or access to one.

Regarding renewable energy systems, despite the support from governments in the form of subsidies, such as grants, lower taxes and lower national insurance for various kinds of renewable energy projects, and to customers willing to convert their systems from fossil fuels to renewable energy, as well as laws and regulations to enforce the approach, renewable energy usage and implementations are still lagging behind. This kind of slow progress is due to a number of factors and challenges faced by renewable energy systems. One of these challenges is the strong competition from the well-established markets and distribution networks for fossil fuels. The opposite is true for renewable energy. However, laws and regulations in support of renewable energy, especially in the West, have increased pressure on various sectors of the energy production industry, such as power generating companies (electricity), vehicle manufacturers, as well as other areas within the oil and gas production: refineries, etc. Despite financial support from governments around the world, providing resources to encourage the implementation, production, and usage of renewable energy, there are still many obstacles facing these technologies. Some of these challenges can be surmised in the following points (Altawell, 2014):

- Financial constraints which limit the greater deployment of renewable technologies, i.e., the barrier lies in the perceived risk associated with investing in renewable energy technologies, which is generally higher than competing in conventional technologies, and the effects of this higher perceived risk in the technology markets.

- Renewable technologies are relatively new in capital markets, and as such, there is more risk than in using established technologies. The higher the perceived risk, the higher the required rate of return demanded on capital.
- The perceived length and difficulty of the permitting process is an additional determinant of risk. The high financing requirements of many renewable energy technologies often present additional cost-recovery risks for which capital markets demand a premium.

In the European Union in general and in the UK in particular, the government provided a number of schemes in support of renewable energy development, such as "The Renewable Obligations Order (2002)," "Feed-in-Tariffs (FITs) (2010)," and "The Renewable Heat Incentive (2011)." These schemes can cover a variety of support, which can be in the form of low-interest loans or loan guarantees, which might serve to reduce perceived investor risk. Tax credits for renewable energy technology production through the early high-risk years of a project may also provide another mechanism for development. Regulatory cost recovery mechanisms, which today often favor low initial cost, fuel-based technologies that can be modified to recognize life-cycle cost as a more appropriate determinant of cost-effectiveness. Effective redistribution of government spending in research and development that directly reflects the potential of renewable energy technologies. The Renewables Obligation Certificate (ROC) born out of the Renewable Obligation Order in 2005 has been defined by OFGEM (the Office of Gas and Electricity Markets) thus:

"A Renewables Obligation Certificate (ROC) is a green certificate issued to an accredited generator for eligible renewable electricity generated within the United Kingdom and supplied to customers within the United Kingdom by a licensed electricity supplier. One ROC is issued for each megawatt hour (MWh) of eligible renewable output generated".

The ROC certificate became law in 2005 when the government issued the Renewable Obligation Order 2005. The ROC obliged the power generating providers a percentage of their electricity produced from renewable sources. According to previous governmental legislation from as early as 2002, every year, the percentage of electricity from renewable sources should be increased e.g., 2006—6%, 2007—7%, and so on, reaching ~20% by 2020. Power generating companies who cannot provide proof (certificates) related to the above could be fined. As a digital certificate, the ROC holds information concerning the production of renewable electricity per unit. The certificate can be traded, as the government guarantees them.

Regulations and various government incentives have resulted in some reduction in CO_2 emissions in the energy production field. This, in turn, resulted in the development of different technologies. At present, there are a number of new and advanced RE technologies, some of which have been passed from the experimental and testing stage into large commercial operations. On the other hand, many new technologies are still in the process of being developed or close to the final testing phase. The use and application of renewable energies are still in the development process, despite a number of factors pushing hard for an alternative source of energy. Currently, the world is experiencing the evaluation process for all types of renewable energy (sources and systems) development before one type or another can take the lead and dominate the commercial energy market - as fossil fuels have been doing since the invention of the combustion engine.

Finally, factors needed for environmentally friendly sustainable fuels, can be summarized in the following eight points (Altawell, 2014):

1. Environmentally friendly
2. Secure and reliable supply
3. Economical — affordable
4. Available on demand (international, national, and local - cities, towns, and villages)
5. Compatible with a wide range of "technology"
6. Grades (produced in different forms, i.e., types)
7. Storage and lifecycle
8. Present and future developments are possible

As final words for this part of the book, a number of questions in a random order have been put forward for the readers to contemplate and answer or provide a particular view.

- **Climate Change and Global Warming**
  o What are the factors related to the causes of climate change? Is it possible for the future generations to survive this kind of environmental changes?
  o What are the factors related to the causes of the global rising of temperature?
  o What is the maximum temperature increase that may still allow the rest of the species to continue surviving in this kind of environment?
  o What are the various issues associated with water security?
- **Energy Storage Systems**
  o Why and how can energy storage systems provide an efficient power supply?
  o What is the relationship between these systems and a Smart Grid system?
  o Can hybrid systems replace energy storage systems?
  o Energy storage is vital for a variety of power generating systems, but especially so for certain types of renewable energy generators, what are they?
- **Crude Oil and Gas**
  o What are the crude oil E&P (exploration and production) approaches taken by oil and gas companies onshore and offshore?

In addition to the challenge of climate change facing humanity, there is another urgent challenge, which can be connected directly, or indirectly to the above issue, this challenge is related to energy and energy shortages, worldwide.

Energy shortages, particularly in the form of electricity supply, are in need of urgent solutions in various parts of the developing world, especially so in the countryside (rural areas).

Finally, the contribution provided in the following pages, in the form of energy (fossil fuels and renewable energy) with related systems and data, covering rural areas/electrifications, directly and indirectly, is what the reader will find in this book.

References

Altawell, N., 2014. The Selection Process of Biomass Materials for the Production of Bio-Fules and Co-firing. Wiley Publisher.

DifferenceBetween.Com, 2011. Difference Between Energy and Force.

Duflo, E., Greenstone, M., Hanna, R., 2008. Indoor Air Pollution, Health and Economic Well-being. http://economics.mit.edu/files/2375. (Accessed 17 July 2020).

EIA, 2010. US Energy Information Administration − India. http://www.eia.doe.gov/cabs/India/Background.html. (Accessed 25 January 2018).

EIA, 2015. US Energy Information Administration − China. https://www.energy.gov/sites/prod/files/2016/04/f30/China_International_Analysis_US.pdf (Accessed 8 July 2016).

IEA, 2007. World Energy Outlook 2007. https://www.iea.org/textbase/nppdf/free/2007/weo_2007.pdf. (Accessed 8 April 2017).

IEA, 2017a. Key World Energy Statistics [1]. https://www.iea.org/publications/freepublications/publication/KeyWorld2017.pdf. IEA, Paris (Accessed 8 April 2018).

IEA, 2017b. World Energy Outlook 2017 [2]. https://www.iea.org/weo2017/. (Accessed 8 April 2018).

IEA, WEO (WEO-IEA), 2017c. World Energy Outlook. https://www.iea.org/reports/world-energy-outlook-2017 (Accessed 17 July 2020).

Kardashev, N., 1984. On the inevitability and the possible structures of supercivilizations. A86-38126 17-88. In: The Search for Extraterrestrial Life: Recent Developments; Proceedings of the Symposium, Boston, MA, June 18−21, 1984. Dordrecht, D. Reidel Publishing Co., pp. 497−504, 1985.

Sachs, J.D., Warner, A.M., 1997. Natural Resource Abundance and Economic Growth. Working paper. Center for International Development and Harvard Institute for International Development.

UNCTAD (United Nations Conference on Trade and Development), 2017. New Innovation Approaches to Support the Implementation of the Sustainable Development Goals. http://unctad.org/en/PublicationsLibrary/dtlstict2017d4_en.pdf. (Accessed 11 April 2018).

World Bank, 2018. World Development Report 2018 Data. https://www.worldbank.org/en/publication/wdr2018/brief/world-development-report-2018-data (Accessed 17 July 2020).

2

Coal

Coal is found in abundance in many parts of the world. It has historically been used as an important source of primary energy, predominately in the 20th Century for the generation of electricity in coal-fired power stations, where the fuel is burnt to raise steam to power-generating turbines. With coal's growth now being supplanted by natural gas, the industry has seen a rise in coal's use in liquefaction processes that produces liquid fuels and gases by the addition of hydrogen molecules to the near-carbon content of coal (see Fig. 2.1).

While coal deposits remain abundant and are predicted to last for more than 200 years at current production rates, the extraction and burning of coal provide health, safety, and environmental issues that are becoming unacceptable risks in today's modern and sensitive society. Burning coal produces emissions that contain sulfurous and nitrogen compounds for acid rain within the earth's atmosphere. That unacceptable issue is in addition to the release of large quantities of carbon dioxide leading to increased rates of global warming, a sensitive issue the world over to governments and their peoples. Coal is, therefore, seen as a "dirty" fuel in the 21st Century, and many governments have encouraged a switch to natural gas—considered to be the cleanest fossil fuel—for power generation and environmental pressure leading to the decline in coal consumption in almost every corner of the globe.

Coal accounts for a near 28% share of global energy consumption (BP, 2018), still bigger than that held by natural gas, which is a cleaner-burning fossil fuel. In BP's analysis of primary energy consumption by fuel type (BP, 2018), coal's actual consumption in 2017

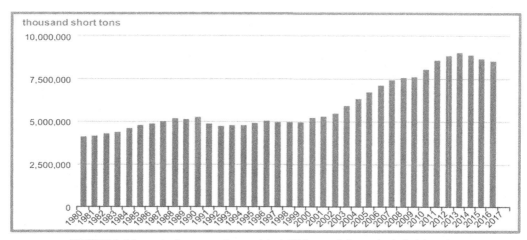

FIGURE 2.1 Primary coal consumption — global. *US Energy Information Administration.*

Rural Electrification. https://doi.org/10.1016/B978-0-12-822403-8.00002-3

was some 3,732 million tonnes oil equivalent, coincidently almost the same volume as it was in 2012 when coal's share of primary energy consumption was 30%. Consumption, considering that coal is the dirtiest of all fossil fuels—is showing no real signs of abating demand.

2.1 Production of coal

2.1.1 The formation of coal

Like all fossil fuels, coal was formed over 350 million years ago by the transformation of organic plant matter. The start point was often a sedimentary basin, on the edges of which swamps formed, and a rise in sea levels, which then covered, and thus killed, the vegetation. This dead vegetation was then covered by sand and mud; the vegetation matter was thus protected from the atmosphere and the rate of decomposition slowed. As sea levels receded, the vegetation grew back until the next cycle of flooding and coverage by mud and sand. This whole process is known as sedimentation.

Eventually, the weight of the deposits causes the sedimentary basin to sink, and the dead vegetation gets cooked by the higher temperatures found nearer to the center of the earth. The heating causes the transformation process through different stages. First, the cellulose in the plants (woody matter) turns from peat into lignite, followed by subbituminous, then bituminous coal. The final stage of this transformation process sees the formation of anthracite, which is almost pure carbon (Box 2.1).

The Carboniferous Period—some 360 to 290 million years ago—contributed the most optimal conditions for the formation of coal, but coal was continuing to be formed since those far-off days. Both the Permian[1] and Mesozoic[2] Periods continued to see coal deposits being formed. More recent coal formations—such as those in the Tertiary Period—are obviously not as mature, and therefore, contain high volatile matter in the

---

BOX 2.1 Insight: types of coal

Several types of coal—a fossil fuel—exist. A normal rank order uses its carbon content and volatility to differentiate these different types.

Anthracite is used for heating and is nearly pure carbon (>85%) with low volatile content (<10%) and thus forms an excellent fuel for heating homes. Next comes **bituminous** coal, used to make coke, and is used in metallurgy. It has a slightly lower carbon content than anthracite (70%–85%) and a higher volatile content (46%–31%). **Sub-bituminous** coal is burned in industrial applications to generate heat and steam. Carbon content is between 70% and 76%, with a volatile matter content of between 53% and 42%. The lowest grade coal used in industrial applications is **lignite**, which has less than 70% carbon content, and a volatile matter between 63% and 53%.

---

[1]Permian Period: 290 to 250 million years ago.
[2]Mesozoic Period: 3250 to 65 million years ago.

Box 2.1   Insight: types of coal—cont'd

Peat is often confused with coal, but strictly speaking, this is not coal, but part-decomposed vegetation matter. It is entirely composed of volatile matter and possesses less than 60% carbon content. Peat is a poor quality fuel that was once used throughout Europe, but its use today is confined to a few regions, such as Ireland. Peat is often formed into briquettes for easier handling.

form of bitumen and decaying wood, and are often lignite. Even more recent accumulations of dead vegetation, which were never buried deep enough to attain any significant level of carbonization, are rich in fibrous plant debris, and likely to have only formed deposits of peat.

2.1.2   Production

Statistics show that, in 2017, global production of coal amounted to some 3,770 million toe[3] (BP, 2018). This was an increase of some 3.2% over the previous year, but some 208 million toes down from the peak from the last decade in 2013, a fall of some 5%.

Production is highly geographical, as China alone accounts for over 46% of global production (BP, 2018), with the USA and Russia producing 10% and 6% of global output, respectively. Indeed, the Asia Pacific region accounts for over 71% of the world's coal production; North America has an 11% share and the Commonwealth of Independent States (CIS) just over a 7% share. The following graphic (Fig. 2.2) demonstrates the geographical distribution of coal production.

About half of the world's coal is extracted from underground mines, the rest from surface mines, which is also known as opencast mining. There are two processes for

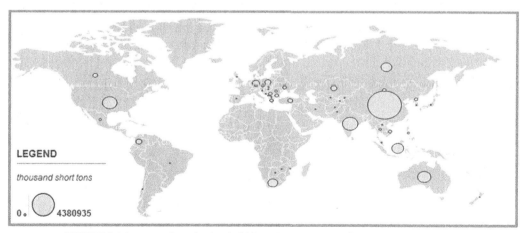

FIGURE 2.2 Primary coal production bubble map. *US Energy Information Administration.*

[3]toe: tonnes of oil equivalent.

underground mining, longwall or "room-and-pillar" (World Coal Association, 2019). Longwall mining uses mechanical shearers to extract the coal from a section of the coal seam. The face of the mining can vary in length between 100 and 350 m. Supports hold up the roof of the mine until all coal is extracted, and then removed, allowing the roof to collapse. This method allows the extraction of three-quarters of the coal deposit, which can extend up to 3 km through the seam.

The term, "Room-and-pillar" is used to describe the extracted coal as the "*rooms*", and unextracted coal as the "*pillars*", as they support the roof. These pillars can represent as much as 40% of the coal in the total seam.

Opencast mining is economical when the coal seam is near the earth's surface. It is more efficient than underground mining, as all the seams of coal can theoretically be extracted. In practice, over 90% of coal is extractable in opencast mining. Opencast mines can extend over large tracts of country, often many square kilometers. Equipment such as draglines, bucket wheel excavators, and conveyors are to be usually found in opencast mining locations. After the overburden of earth and rock is removed, then the exposed seams are drilled and fractured to enable mining to take place in sections before removal from the opencast site.

With environmental pressures nowadays, worked-out opencast mines are restored and rehabilitated, so that no evidence of the mining activity is left behind.

2.1.3 Exploration

Total proved reserves in 2017 amounted to some 1,035,012 million tonnes, of which around 70% was anthracite and bituminous quality. These reserves are split almost 50/50 between OCED and non-OECD countries. According to the *Statistical Review of World Energy* (BP, 2018), at current rates of production, reserves can last for 134 years. Reserves that are classified as proven are those that geological and engineering opinion considers to being recovered in the future with reasonable certainty under existing economic and operating conditions. It should be noted that not all these huge reserves will ever be consumed. Many reserves have no economic value, and the US shale oil and gas explosion has also eroded much of the value of such coal reserves. The following world map illustrates the distribution of coal reserves (Fig. 2.3).

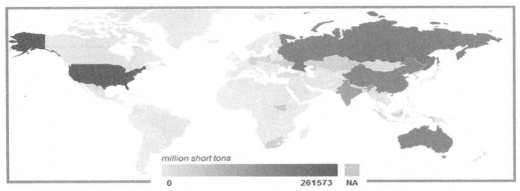

million short tons

0   261573   NA

FIGURE 2.3 Coal: recoverable reserves by country. *US Energy Information Administration.*

2.1.4   Geography

Three out of the top six coal producers are the world's biggest consumers. Of these three, first is China, which produces 46% of global coal production, and consumes around 51% share of the global consumption. Then comes India, with around a 10% production share, but an 11% share of consumption. The USA is the world's third-biggest producer, but its consumption and production shares are balanced at 9% of the global figures. The other three largest producers are Australia, Indonesia, and Russia, all with a 5%–6% share of global production. Australia and Indonesia are the biggest exporters of coal, with 391 and 369 million tonnes per annum. Despite its large production, China is the world's biggest importer with over a quarter of a million tonnes of imports each year.

2.1.5   Geopolitics

In the light of the world's focus on climate change and greenhouse gas emissions, any country building a new coal-fired power station faces condemnation and hostility from environmentalists. Investors and banks no longer finance coal-fired power plants. Historical US coal mining giants such as Peabody and Arch Coal have fallen to bankruptcy in the past decade. As just one example, US coal production fell by nearly 40% in just 1 year as of April 2016.

Emerging economies such as India and those in southeast Asia will continue to drive up demand for coal imports, but this increase is unlikely to fully supplant China's waning demand.

2.1.5.1   Industry structure

The coal mining world is highly fragmented, with many organizations contributing to market share. Some of the world's largest and most successful coal mining companies are briefly described below.

The single largest coal miner globally is **Coal India,** a state-owned organization with some 80+ mining areas, which mined over half-a-billion tonnes in 2016. Driven by India's political desire to achieve improved energy security for its burgeoning population, Coal India is seeking to acquire foreign assets, with coking coal sources in Australia and other South African targets for its acquisition policies.

BHP Billiton can trace its ancestry back to Indonesia in the 19th Century. Billiton was a tin miner, but when merged with BHP—a minerals mining company from Australia—it has become one of the global conglomerates. As well as coal, BHP has interests in copper and iron ore, as well as petroleum. More recently, cost reductions, and greater productivity initiatives have resulted in all its operations becoming positive cash generators, thus anticipating improved future market conditions.

By contrast, a much more modern creation, China's **Shenhua Group**, is a directly state-owned and controlled Chinese company that was started in 1995. It is the most prestigious coal miner in China, with high levels of modernity and the greatest global distribution reach. A recent merger of Shenhua with power-generator Guodian Corporation has created a $271 billion behemoth to become the world's second-largest company by revenue.

After a debilitating filing for bankruptcy protection, the US-based **Arch Coal** has become one of America's leading coal producers, selling 96 million tonnes of coal in 2016. Active coal mining operations in Wyoming, West Virginia, Colorado, and Illinois represent a 13% share of America's coal demand.

Anglo American plc can trace their roots back over a 100 years to the gold and diamond fields of South Africa. Anglo American has evolved into a large diversified miner, with a geographically diverse coal mining portfolio of assets. They are the world's third-largest supplier of metallurgical coal and are diversifying by selling off non-core assets to sharpen its focus to meet the future challenges facing all the companies involved in coal mining with the transition to a low-carbon world.

2.2 Fiscal systems

Each country will issue its own regulations and processes, as well as enact laws to govern the exploration and production of coal resources; in addition, these rules may extend as to the level of permittable exports, satisfying domestic demand and benchmarking coal prices.

The Indonesian government, through the Minister, sets a benchmark price on a FOB basis vessel point of sale, and this benchmark price serves as the floor price in royalty calculations if the actual selling price of the coal is below the benchmark. In situations where the actual selling price is above the benchmark, then it is the higher actual price achieved by the trade coal that is used for royalty calculations.

Provided the regulations are policed efficiently by the government's ministries, the opportunity for windfall profits are mitigated; the fiscal regime is designed if not to favor the host government, then at least to save them from being exploited by unscrupulous operators (Box 2.2).

Traditional mining companies sell their output on a trading basis as a commodity. There is little vertical integration between the coal miners and the power generation utility firms. Any liquidity in the coal trading market comes from these utility companies only.

The coal derivatives market is based on three major trading hubs, Newcastle (Australia), Richards Bay (South Arica), and Rotterdam in Europe. The Australian market is based on a derivative developed by Global Coal based on physical trades, with no market assumptions; coal is traded on a FOB[4] Newcastle basis. The South African derivative is known as "AP14" and is a standardized contract for 150,000 tonnes of physical coal FOB Richards Bay. The APA12 contract is on a similar basis but for physical standard steam coal delivered into ARA.[5] Both latter derivatives use market assessments provided by Petroleum Argus and McCloskey.

[4]FOB: Free on Board indicates that the buyer is responsible for all costs and risks once the coal has been loaded onto the ship at the port specified; in this case, Newcastle. Australia.

[5]Industry acronym for the ports of Amsterdam,. Rotterdam, Antwerp and surrounding ports within that area.

> **BOX 2.2 Insight: example fiscal system — Indonesia**
>
> As an example of the complexity of these issues, we shall use Indonesia as an example from a report written on investment and taxation in the Indonesian coal sector (PwC, 2016). Indonesia's fiscal system is contained within the provisions of the Coal Contract to Work (CCoW). Companies wishing to extract coal must obtain a license to explore (usually for up to 7 years), followed by a separate license to produce (usually 20 years). Foreign companies are required to offer shares up for Indonesian ownership, such that after 10 years, foreign participation is reduced to 49%. Royalties are to be paid, with their precise rate dependent upon the scale of mining, production levels, and the price of coal. Typically, these were 13.5% calculated from the selling price of coal less specified marketing costs. Land rents for a coal mine in production ranged from $1.00 to $4.00 per hectare per annum; lower rents applied for survey, exploration, and feasibility phases. Corporate income tax of 35% applies for the first 10 years, rising to 45%. Other rules strictly govern depreciation and amortization rates, allowable expenses, investment allowances, and so on.

2.3 The global coal market

Coal consumption represents over a quarter of total primary energy consumed worldwide, and it is still the largest source of power generation at 38%, according to BP Statistics 2018. China, the USA, India, Australia, and Indonesia are the largest global coal-producing countries; these countries are also the largest coal consumers. Most of the global coal production is consumed in the country in which it is produced, with only 15% of total production directed to international markets.

Coal has experienced a great surge in the early 2000s, mainly due to the exceptional growth of the Chinese economy. On the other hand, demand for coal in OECD countries started to weaken and by the second decade of 21st Century experienced a significant decline. Coal still plays a key role in economic development for the largest coal producers. Access to affordable and secure energy is the propeller of economic development and a key issue for Asian countries such as India and China experiencing rapid industrialization and urbanization. For these developing countries, the environmental agenda has to be reconciled with imperatives of access to energy, energy security, and economic development. These countries rely on coal to meet a great proportion of their energy needs and are likely to continue to rely on coal in the future. According to the World Coal Association, 20% of the population representing 120 million of people still live without electricity in Asia, and the South East Asia (ASEA) region, and 45% rely on fuel-wood and charcoal for cooking and heating.

According to the International Energy Agency (IEA), the coal market will continue to expand in China and India, offsetting some of the declines in Europe and the US; however, the growth of demand in China will be more moderate. Globally, the share of coal in the energy mix is expected to decline from the current 27% to 25% in 2023 (IEA, 2018).

In spite of the growth in global demand, coal is attracting less new investment due to global climate change policies and the perceived high risk of investment in the sector. Western European countries have adopted clear phase-out plans for coal. China is also adopting environmental policies driven by concerns over air quality, which is expected to have a detrimental impact on the prospects of growth in coal demand.

Some Asian countries have signed the Paris Agreement committing to end unabated coal power generation by 2030, but some other countries in the region are still relying on coal for affordable access to energy. The prospect of continuing growth in coal in Asian countries is colliding with the environmental commitment to reduce emissions. A solution such as pollution abatement technology and carbon capture and storage (CCS) could be part of the solution. However, these technologies are still not sufficiently advanced to become commercially viable, and more support is required in terms of research, development, and finance to enable progress and advance these technological solutions.

2.3.1   Supply and demand

BP's *Statistical Review of World Energy* (2019) indicates that world coal consumption grew in 2018 by 1.4% after several years of decline, and this was attributed to strong growth from Asia, particularly from India and Turkey, with China showing only a slight increase. Production of coal in 2018 surged by a 4% growth - quoted by BP statistics - mainly from increased coal production in China, India, and Indonesia and also from other regions of the globe. In spite of the slight upturn in levels of world consumption and production in 2017 and 2018, coal's share in the global primary energy mix fell to the lowest level at 27.2% in 2018 following decades of decline (Fig. 2.4).

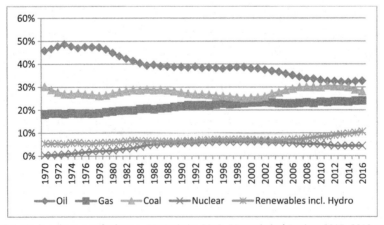

FIGURE 2.4 Market share per fuel type. *BP statistics 2018, BR statistical Review 2017, 2018 and 2019.*

Coal consumption has been declining in OECD countries - this is mainly competitive pressures from lower gas prices and environmental pressures to switch to cleaner energy due to sources. In OECD countries, the contribution of coal for power generation has declined from 38.6% in the year 2000 to 27% in 2018. (IEA, Electricity Information, 2019). In contrast, for non-OECD countries, the contribution of coal to power generation stands at 46.5%.

China and India are the first and second-largest producers of coal and its largest consumers. China and India are heavily dependent on coal in their overall energy consumption of 58% and 57% respectively in 2018. Dependency on coal for power generation is even higher at 70% for China and 72% for India, according to the Energy Information Administration (EIA).

China dominates world coal markets, accounting for more than half of total global demand. The EIA believes that Chinese coal consumption may now be weakening, with industrial use for steam and steelmaking already having peaked, and coal demand for power generation likely to peak around 2023. Taking into account that demand is also in long-term decline in the US and in Europe, growth in some emerging economies, led by India, may not enough to raise overall coal use.

In Europe, there are significant differences in policy approaches towards coal. While Western Europe experienced consistent declines in coal consumption propelled by climate change action and coal phase-out plans, coal-fired power stations are still being built in Eastern Europe. In Western Europe, the consumption of coal has been declining for the last 2 decades and accelerated since the turn of the 21st Century. The use of coal for power generation has experienced a steep decline being replaced in great part by gas and renewable energy. In 2018 alone, coal generation declined by 6% and has cumulatively dropped 30% below 2012 levels. Environmental factors are driving the switch away from coal; the EU is committing to cut emissions from the 1990 levels by 40% by 2040.

Three-quarters of the hard coal for generation in the European Union are to be subjected to national phase-out plans, which will take place in the next 10 years up to 2030, with the exception of Germany, which announced the phase-out only by 2038 (Agora Energiewende, 2018). Coal power generation is expected to fall below gas-fired power generation by 2020. Both gas and renewable energy are continuing to erode coal's market share in power generation. The application of environmental energy policy is likely to have a significant impact, and the resulting rise in carbon price will further undermine coal's economic viability.

In Europe, however, the question of gas imports has raised security issues, and the region has taken steps in terms of diversifying its gas suppliers away from the substantial reliance on Russian gas supplies. While the production of coal in European countries has been declining, there has been an increase in imports into the region since the late 1990s. These imports are not expected to increase as European switches to gas, renewables, and nuclear to fill the demand requirements.

Both China (70%) and India (72%) maintain a significant dependency on coal for electricity generation. Access to energy and electricity, in particular, are key aspects of the economic development effort and contributing to lifting their Human Development index, narrowing the gap to the Western countries. Both countries have been increasing their coal imports to supply the ever-increasing demand for electricity. China's coal reserve represents 13% of global coal reserves, but it consumes 51% of the world's coal demand.

In the US, coal production represents nearly 14% of the total energy mix, but both production and consumption have experienced substantial declines since the start of the 21st Century (EIA, Monthly Energy Review, April 2018).

US Investment in coal has been challenged by robust competition from gas and renewable energy - the switch to gas has been facilitated by good gas economics due to the abundant availability of shale gas.

In the US, coal still plays a pivotal role in electricity generation accounting for 30% of total power generation in 2018, but down from 45% in 2009. Most of the decline in coal consumption is explained by the competitive challenges from cheap gas and renewable energy resulting in reduction in capacity utilization and early retirement of coal-fired power stations. (EIA, Monthly Energy Review, October 2018).

2.3.2   Global trade

China and India will continue to be key market players while Europe was estimated to become a marginal coal market accounting for only 6% of global consumption in 2018.

The IEA is expecting global trade in coal to decrease after 2020 due to climate change pressures and domestic policies encouraging the development of cleaner energy resources.

To supply their increasing electricity demand, China and India have been increasing their coal imports. China consumes more than half of the world's coal consumption and recourse to coal imports accounts for a substantial part of China's energy requirements. However, while China is still pursuing coal investment to supply energy needs, there are parallel efforts to decarbonize the economy with substantial investment in renewables energy both in the power generation and transport sectors. One of the policy drivers is to improve air quality and reduce smog in the Chinese cities by reducing the incremental growth in energy demand coming from coal. The decarbonization efforts have been helped by a reduction in costs of producing power from renewable energy. India continues to up its coal imports as domestic coal does not meet the calorific value required for power generation.

Australia became the largest coal exporter in the 1980s. An increasing proportion of the exports of Australia steam coal found its way to Asian markets displacing some of the hard coking coal from the US. Australia's market share of world trade has gradually expanded from 17% in 1997 to near 38% estimated for 2020 and is continuing to supply half of the coal imports into the Asian continent (Shepherd and Shepherd, 2014). Australia is the fourth-largest producer of coal behind China, US, and Indonesia, representing only 6.9% of world coal production, but it has become the largest exporter in the world exporting around 90% of its country's total production. (IEA, Key World Energy Statistics, 2019).

After several years of decline, US exports of coal started to increase in 2017 and 2018, and exports to Asia doubled in 2017 compared to previous years. This significant increase in US coal exports was due to the high demand for steam coal from Asia combined with supply disruption from Australia and Indonesia, which are the traditional suppliers of coal to Asian countries. (EIA, Quarterly Coal Report, April 2018). The modest increase in US coal production in 2017, together with the decrease in domestic coal consumption, created extra coal availability for export.

India, South Korea, and Japan were the main markets for steam coal, with India being the largest recipient due to rapid growth in electricity demand (EIA, Quarterly Coal Report, April 2018). In spite of being a considerable coal producer, India does not produce the quality of coal suitable for power generation, and it has to turn to imports from other countries.

Japan and South Korean have also been key markets for US coal exports. Japan has increased investment in coal-fired power stations, reflecting clear reluctance to restart nuclear stations following the Fukushima Daiichi nuclear accident. South Korea has chosen to move away from nuclear power and started to increase investment in coal-fired power stations. As a result, these two countries have experienced significant increases in coal imports for power generation.

2.3.3   Characteristics of coal

Coal can be grouped into four different types: peat, lignite, bituminous, and anthracite. Peat is of limited value and use, while the other three fulfill specific roles. The latter is of considerable superior value due to high heat content. A practical approach to coal focuses on the heat content and carbon content. According to Speight (2013) anthracite has the highest concentration of carbon (86%−98% w/w), and it has the highest heat value (13,500−15,600 BTU/lb). Lignite is the poorest of the three coals with the lowest heat value and carbon content.

Coal has multiple applications, from power generation to steel production. Steam coal is the high-quality coal grade used in electricity generation while 'coking' coal is used for steel, iron, or cement making. In steel production, the coke is combined with ore, iron, and limestone to produce steel. Coal can also be used as a chemical source for industrial production of various synthetic products (dyes, waxes, pesticides and so on).

Coal is used for power generation and commercial heat accounting for 65% of overall coal use in the world, reaching 85% in OECD countries. This reflects the almost complete switch from the use of coal for non-power generation purposes.

Coking coal is used in manufacturing, especially in the production of pig iron. Coal has an important role in steel production as coking coal is a crucial part of blast furnace steel production, which contributes to 70% of overall steel production. Steel is the touchstone in modern lifestyle, being the essential material in both private and commercial domains, from transport to telecommunications and from agricultural to industrial manufacturing products.

In recent years China has become the largest steel producer with a global market share of 49% in 2014 compared to only 15% in 2000. Lower manufacturing costs have allowed the country to climb to the first spot in world producers, followed by Japan, the United States, India, and Russia. In developing countries, steel production is expected to grow as a necessary condition for the development of vital infrastructure; this opens up the prospect for substantial growth in the coking coal sector (World Energy Council, 2016).

2.3.4   Coal value & prices

The value and price of coal is determined by its heat and carbon content. The value of coal is also a reflection of mining costs as underground, and more difficult to extract, coal has higher costs than the surface coal deposits. The price is also influenced by environmental regulations; therefore, low sulfur coals tend to exhibit a higher price than high sulfur coals. Prices are normally higher for Anthracite coal, which has higher heat content - it is a hard black colored coal with a carbon content between 86% and 97%. Anthracite is the more mature and rarest of all the coal types, and it is more commonly used in metallurgical and other industrial processes.

The second highest quality is bituminous with a carbon content between 77% and 87% on a dry ash-free basis percentage. There are two types of bituminous coal: one to be used as thermal fuel for power generation and the other for metallurgical and cement industrial production.

Sub-bituminous ranges from black to dark brown and ranks between anthracite and lignite; it is used primarily for steam-electric power generation but can be used for industrial purposes; it contains between 70% and 77% carbon on a dry ash-free basis (Geology.com).

Lignite is the lowest ranked coal with relatively low heat content and the lowest fixed carbon content of between 60% and 70% (Geology.com). It is often referred to as brown coal and soft in nature derived from sedimentary rock formed naturally from peat. Most of the lignite is used for power generation.

Coal is priced according to different types in the market. Coal benchmarks reflect the different coal characteristics and these benchmark prices are the basis for setting future contracts in global exchange markets. The price of coal is influenced by multiple factors, the main ones being: supply and demand, price of alternative energy sources and environmental policies. If alternative fuels have more attractive prices, the demand for coal tends to drop and vice-versa.

Prices for the main coal grades in 2017 quoted by EIA:

- Bituminous—$55.60
- Subbituminous—$14.29
- Lignite—$19.51
- Anthracite—$93.17
- Prices are quoted as dollars per short ton (2,000 pounds)

Price is also affected by freight costs as a large amount of coal is traded internationally. Supply will affect prices significantly as oversupply will tend to depress prices. In recent years, cheap gas prices have depressed coal prices and made coal less competitive versus alternative gas energy sources. Most of the coal demand is derived from the power sectors. Increasing demand from developing countries of Asia was triggered by higher economic activity and high economic growth. Rapid industrialization in those countries has increased demand for steel, which requires coal, leading to an increased demand for coal.

Price volatility is here to stay. Changes in China, whether in policy or economic circumstances, feed volatility in global coal markets, given its sheer size and dominance in the global coal trade. When combined with supply disruptions, this volatility is amplified. Prices will continue to depend largely on China; as a consequence, the structural reform of the Chinese coal industry is key in the evolution of coal prices.

Among exporting countries, Indonesia deserves special attention: expanding domestic demand combined with constraints on ramping up production might increase market tightness and push prices up. On the demand side, import levels from China, India, Korea, and Japan are key uncertainties. However, falling demand from OECD countries and slow demand from China may have a significant impact on prices in the long term.

US coal has suffered highly competitive pressures from alternative producer centers such as South Africa, Australia, Venezuela, and Columbia. These countries have been able to expand market share due to aggressive pricing strategies in order to increase competitive advantage, and this was also supported by favorable currency movements.

2.3.5   Forecast for the coal market

The IEA is estimating that global coal demand will remain largely flat, with consumption levels by 2023 being similar to the 5-year average for the period between 2013 and 2018. Coal share in the global energy mix is expected to decrease from 27% in 2018% to 25% by 2023 due to slower demand. In the power generation sector, the contribution of coal is expected to drop to 36% from a historical average contribution of 40% in spite of overall increase in coal-fired coal production for the period up to 2020 (IEA, Coal Information, 2018).

Climate change policies in OECD countries have defined the agenda for switching to cleaner energy sources. In China, issues of air quality have prompted a slight change in policy and more investment in renewable energy. The growth in consumption of coal is moderating and even expected to become flat. While in India, consumption of coal is still strong but moderating from a high level of 6% growth per year in the last decade to an average of 4% growth up to 2023. (IEA, Coal Information, 2018). According to the IEA, coal consumption in the EU will drop 2.5% per year from 325 tonnes of coal equivalent in 2017 to 280 million tonnes equivalent in 2023.

Coal is estimated to continue to be an essential source of energy in China even when considering the structural changes taking place in the Chinese economy, including the move from heavy industrial energy-intensive industries to a more service-oriented economy and efficiencies reached within the manufacturing industry and the power sector resulting from shutting down of old facilities and the move toward investment in less carbon intensity industries. In China, issues of air quality have prompted a slight change in policy and more investment in renewable energy.

According to some commentators, the US will fulfill the role of swing producer in international markets likely to take advantage of supply disruption from Australia and favorable high coal prices, which would make freight economically feasible for US coal exports to Asia.

BP's forecast for coal points to near-zero growth for the period 2010 to 2030 and slow decline after 2040 for OECD countries. Consumption in the US and Europe is continuing to decrease with forecasts estimating that the trend will continue indefinitely as countries in the West are switching from coal-fired power generation to gas and renewable energy; this trend being driven by environmental commitments.

2.3.6   Is coal supporting rural electrification?

Development in OECD countries was in the past heavily anchored on the use of coal, but consumption in these countries is continuing to switch away from coal and into gas and renewable energy. Access to energy is considered to be the cornerstone of economic development and promotion of living standards; access to energy is also a central part of the UN Sustainable Development Goals (SDGs). Demand for electricity grows as urbanization and industrialization efforts intensify in developing countries. However, in rural areas, the problem of energy access is more acute, and the connection between access to energy and development efforts is not automatic and must be supported by government initiatives to promote rural income growth. This could take the shape of provision of basic infrastructures, such as rural water supplies, health, and education programs, and feeder roads. No access or inadequate access to electric power in rural areas means there is a dependency on coal, charcoal, or biomass to fulfill the basic needs of cooking and heating. At the same time, this limits the scope for developing economic activities and also income generation, trapping the rural population into a permanent cycle of poverty. Studies by the World Bank indicate that access to electricity associated with agricultural development in rural areas are key factors in poverty reduction. (World Bank, 2017). Energy affordability is a fundamental aspect to support improved agricultural productivity and a step toward more sustainable economic development.

For developing countries, electrification produces some challenges, and these countries are presented with diverse options, often between the use of domestic resources, such as coal or other energy sources such as renewable energy. For countries such as sub-Saharan Africa, the challenges are immense, taking into account the task to be undertaken. There are, however, some success stories in the region - South Africa has evolved from an electrification cover of 36% in 1994 to 75% in 2009, and 86% in 2015. South Africa, as a major coal producer, uses coal to generate most of its electrification requirements because of its availability and affordability. Cost is, without doubt, dependent on local resources, and particularly, circumstances. Traditionally coal has been cheaper in China compared to gas or oil. This is because coal is cheap and available domestically. In other countries in Asia, for example, Malaysia and Vietnam, gas is cheaper than coal as gas is produced in those countries and piped to customers.

In developing countries, the focus of electrification has been directed toward urban areas, with rural regions being largely marginalized. Tackling electrification in rural areas has been considered an uphill task due to financial and logistical constraints. Also, low-income populations and dispersion were identified as significant obstacles to project implementation in rural areas. The dispersed nature of the rural population in sub-Saharan Africa makes the electrification costs of grid transmission and distribution infrastructure too high to be economically viable. This is one of the main reasons why off-grid options appear to be more popular together with more recent environmental arguments. Small hydropower projects have also been popular in the last 20 years. Small hydropower technology has been extensively utilized in China to address electrification requirements with installed capacity accounting for 51% of the world's small hydro-power (23.5 GW). Photovoltaic technology is considered to be more cost-effective in rural areas of low-density populations with low and infrequent demand levels. The African continent appears to be well-suited for utilization of this technology - Egypt, South Africa, and Kenya are the countries, which are presenting the highest application of photovoltaic systems. This has been possible by the implementation of innovative charging systems based on the payment of monthly rent for the leasing and use of equipment; some of these financial models are affordable to customers but deliver a return to investors.

A recent study by Bazilian et al. (2011) indicates that to fulfill power demand in developing countries in sub-Saharan Africa will require production capacity, 13 times current provisions and this incremental capacity to be deployed within a 20-year time scale - this would require, according to this study, an electrification growth rate of 12% per year, which contrasts with a modest production capacity average growth rate of 1.7% per year for the last 20 years.

The energy investment options presented to developing countries involve a careful analysis of available energy resources and the assessment of local socio-economic circumstances. In the face of environmental and climate change pressures, it is interesting to analyze if expansion in electrification in developing countries can be directly connected with the growth in consumption of coal or the adoption of more environmentally friendly alternatives.

A study by Yang et al. (2019) from Columbia University analyzed 122 developing countries around the world to find out if the growth in electrification was associated with growth in consumption of coal differentiating between urban and rural areas. Their findings indicated that there is very little connection between rural electrification and growth in coal-fired power stations.

According to Barnes (2007), the decision-making involved in the choice of adequate energy sources is a dynamic process. Factors such as availability of energy resources, population density, demographics, ease of access, socio-cultural, and political realities appear to be key forces in the decision-making process.

In rural and difficult-to-access locations, off-grid or micro-grid systems appear to offer cost-effective power options. The various sources of energy for these systems have their strong and weak points. Diesel generators are easy to install, but fuel and parts can be costly to purchase and maintain. Micro or small hydropower are easy to maintain but require heavy capital investments. Renewable energy, particularly solar power, can be quite appropriate in sub-Saharan countries of Africa, which benefit from the necessary natural resources. Solar PV projects have been successfully implemented in developing countries in Asia, South America, and Africa. Studies conducted by the United Nations indicate that involving the population in energy access projects plays a crucial part in the success of the project with programs of education and training attached to it. Institutional support and some subsidies by government allied with micro-finance or capital grants are key in electricity connections together with the purchase of clean stoves to replace highly polluting traditional stoves using unprocessed biomass or coal.

2.4 The future of coal

Strangely, coal staged a mini-revival in 2017, with both consumption and production increasing, albeit in single-figure percentages. Both India and China recorded increases, and not just from power generation. Coal-to-gas fuel switching obviously has a depressive effect on coal consumption, but short-term spikes in demand from both domestic and industrial users see coal being used as a balancing fuel supply, especially in China.

Climate change measures target coal as the worst polluter of fossil fuels, but some countries are coal-rich and oil-poor, such as China, and to feed their GDP growth and the consequential energy demand, it becomes difficult for politicians and policy-makers not to exploit their natural resources. GDP growth, driven by increasing industrialization, can overcome carbon mitigation measures, such that global carbon emissions continue to grow despite quite severe carbon reduction policies in other developed economies.

There have been calls to phase out the use of coal by 2030 from low-lying Pacific Ocean countries, who face the prospect of rising sea levels, brought about by increases in global temperatures as a direct result of global warming. Some countries are already well advanced in phasing out the use of coal in power stations for electricity generation, notably Germany and the UK.

Coal-fired power stations emit about twice the amount of carbon dioxide than that generated in gas-fired plants. A coal-fired station may generate around a tonne of CO_2 for every megawatt-hour of electricity generated. A decline in indigenously mined coal in a country may be accelerated by the availability of cheaper imported coal, as was the case with the UK, which also saw a "dash for gas" during the 1990s when newly privatized electricity companies were able to take advantage of regulatory change that allowed the use of natural gas in electricity-generating power stations. These gas turbine stations were cheaper and quicker to build than either coal-fired or nuclear power stations. The graphic (see Fig. 2.5) shows CO_2 emissions from coal by country.

As has been already identified, reserves of coal are abundant, but consumption in many countries has irreversibly declined. In the main, coal-fired electricity generation has been replaced by renewable sources or natural gas. It was felt that peak coal production had been reached in 2013, but the small upturn in production and consumption seen in 2017 has put a hold on that conclusion as definite.

Coal also comes under pressure from the reducing costs of nonfossil fuel energy sources such as renewables, solar, and battery. Coal does, however, remain the world's cheapest energy source, at least in some regions of the world; so even though environmentalists are waging a "*war on coal*" on many fronts, in the short-term at least, coal will continue to contribute significantly to the energy mix in many economies across the globe.

2.5 Transition to rural electricity — is this feasible?

Coal is still being burned in domestic premises, either to provide the primary source of heating or as a feature in traditional fireplaces. Over a number of years, coal burnt in such places has become smokeless to reduce emissions of particulates, which in the city and urban areas contributed to smog atmospheric conditions, that still exist today in some places, especially China.

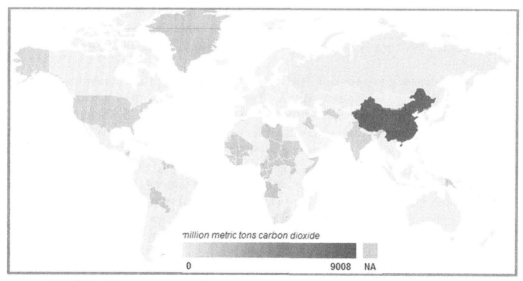

million metric tons carbon dioxide

0   9008   NA

FIGURE 2.5 Global CO_2 emissions from coal by country. *US Energy Information Administration.*

As a source of carbon emissions, coal and charcoal can be easily supplanted by electricity to provide heat, light, and power. However, if rural areas lack the basics of electricity, then invariably, people will turn to fossil fuels for lighting (kerosene) and heating (coal); therefore, any rural electrification proposal should result in immediate benefits in the reduction in carbon emissions and decreasing reliance on fossil fuels.

2.6 Conclusion

While coal founded the industrial revolution and provided the basic foundation for 21st Century globalization, it faces increased challenges from a global desire to de-carbonize our world. Just 10 years ago, coal-fired power stations were generating more than half of America's electricity. Aside from the attack on the industry to greatly diminish carbon emissions, coal has also faced significant competition from natural gas as a primary energy source. Easier transportation across vast oceans, of liquefied natural gas linking distant markets with production wells, has dramatically eaten into what was once called "King Coal."

Further losses to the rise of renewable energy such as wind and solar, would appear to have been the last death knell of this industry but remarkably coal has undergone a small revival since 2017 with countries such as China, India, and the US, supporting the use of coal as a primary energy source. An industry blog (Technavio, 2019) predicts that the coal market will grow by just under 2.0% a year over the next 4 years.

In BP's Statistical Review of World Energy (BP, 2019), this growth is reported, along with the dichotomy of an increase in coal production of 4.3% in 2018, yet at the same time, coal's share of primary energy sources fell to its lowest in 15 years and stood at 27.2% share. Increases in both consumption and demand in 2018 were at their fastest rates for some 5 years. Asia continues to underpin this strength—another couple of years with these levels of increase maintained would see coal back to 2013 consumption levels—a figure which many commentators believed signaled the ultimate peak in coal consumption.

So perhaps coal is not yet dead? The will to decarbonize in theory cannot overcome the increasing energy, and electricity generation, demand from a world where the population growth and subsequent energy demand cannot be met by incremental gains from renewables alone. It is ironic to think that coal, in the twilight of its years, may yet be a balancing fuel, despite its carbon content being completely at odds with a low-carbon world and the hopes and aspirations of politicians and policymakers.

The global coal market will continue to be heavily influenced by demand from Asian countries, particularly India and China. However, a significant level of uncertainty is affecting the coal market. Some of these uncertainties are related to Chinese policies, and the government attempts to balance environmental pressures with economic support for the domestic coal industry. Nevertheless, the uncertainties will add some level of volatility to prices. However, strong demand for thermal coal from India is expected to bring considerable support to coal prices, as the new capacity of coal-fired power comes on line. The projected global slowing down in economic growth could add to the uncertainty and lead to the weakening of coal demand. The uncertainties are also deterring investment in new coal capacity with banks and equity investors perceiving investment risks to be high in light of economic uncertainties and climate change priorities.

References

Agora, 2018. European Power Sector in 2018. https://www.agora-energiewende.de/fileadmin2/Projekte/2018/EU-Jahresauswertung_2019/Agora-Energiewende_European-Power-Sector-2018_WEB.pdf [Online]. Available at: https://www.agora-energiewende.de/fileadmin2/Projekte/2018/EU-Jahresauswertung_2019/Agora-Energiewende_European-Power-Sector-2018_WEB.pdf.

Barnes, D.F., 2007. The Challenge of Rural Electrification: Strategies for Developing Countries. Resources for the Future, Washington, DC. Available at: Google Scholar.

Bazilian, et al., 2011. Energy Access Scenarios to 2030 for the Power Sector in Sub-saharan Africa. Accessed in October 2019. http://services.bepress.com/feem.

BP statistics 2018, BR statistical Review 2017, 2018 and 2019.

BP, 2018. *BP Statistical Review of World Energy*. BP, London.

BP, 2019. Statitical Review of World Energy 2019. BP, London.

EIA, 2018. Monthly Energy Review. April 2018. [Online]. Available at. https://www.eia.gov (Accessed April 2018).

EIA, 2018. Monthly Energy Review. October 2018. [Online]. Available at: https://www. eia.gov (Accessed October 2018).

EIA, 2018. Quarterly Coal Report. April 2018. [Online]. Available at: https://www. eia.gov (Accessed April 2018).

IEA, 2018. Analysis and forecasts to 2023 [Online]. Available at: https://www.iea.org/reports/coal-2018 (Accessed February 2019).

IEA, 2018. Coal Information 2018 [Online]. Available at: https://www.iea.org/reports/coal-2018 (Accessed February 2018).

IEA, 2019. Key World Energy Statistics, 2018. [Online]. Available at: https://www.iea.org (Accessed September 2019).

IEA, 2019. Electricity Information 2019. OECD Publishing, Paris [Online]. Available at: https://www.oecd-ilibrary.org/energy/electricity-information-2019_e0ebb7e9-en.

PwC, 2016. Mining in Indonesia, Indonesia. Price Waterhouse Cooper.

Shepherd, W., Shepherd, D.W., 2014. Energy Studies, 3rd Edition.

Speight, J.G., 2013. The Technology and Chemistry of Coal. CRC PressComposition of Coal [Online]. Available at: https://geology.com/rocks/coal.shtml [Accessed in February 2019].

Technavio, 2019. Technavio Blog [Online]. Available at: https://blog.technavio.com. (Accessed June 2019).

World Bank, 2017. State of Electricity Access Report 2017. Available at: http://documents1.worldbank.org/curated/en/364571494517675149/text/114841-REVISED-JUNE12-FINAL-SEAR-web-REV-optimized.txt.

World Coal Association, 2019. [Online] Available at: ww.wordcoal.org/coal/coal-mining. (Accessed 24 February 2019).

World Energy Council, 2016. World Energy Resources 2016. World Energy CouncilAvailable at: https://www.worldenergy.org/assets/images/imported/2016/10/World-Energy-Resources-Full-report-2016.10.03.pdf.

Yang, J., Urpelainen, J., Hopkins, J., 2019. Is coal-fired power generation associated with rural electrification? A global analysis. Energy Res. Soc. Sci. J. [Online]. Available at: https://papers.ssrn.com/sol3/papers.cfm?abstract_id=3442522.

Further reading

Africa electrification rates, n.d. Accessed in October 2019 at: http://www.unesco.org/new/fileadmin/MULTIMEDIA/HQ/SC/temp/wwap_pdf/Table_Access_to_electricity.pdf.

Canadian Energy Research Institute, 2017. Canadian Oil Sands Supply Costs Abnd Development Projects (2016—20136. CERI, Calgary, Alberta.

China small hydro, n.d. http://hydroint.zek.at/hydro-international/news/chinas-leading-role-in-small-hydro-development/. (Accessed in October 2019).

Coal Market Outlook. [Online]. Available at: https://www.refinitiv.com/perspectives/market-insights/coal-market-outlook-2019/. (Accessed in February 2019).

Coal Price and FX forecast. [Online]. Available at: https://home.kpmg/content/dam/kpmg/au/pdf/2019/coal-price-fx-consensus-forecast-december-2018-january-2019.pdf. (Accessed in February 2019).

EIA, 2018. International Energy Outlook 2018 [Online]. Available at: https://www.eia.gov (Accessed December 2018).

EIA, 2019. Short term Energy Outlook. February 2019. [Online]. Available at: https://www. eia.gov (Accessed February 2019).

IEA, 2019. Coal Information 2019 [Online]. Available at: https://www.iea.org/reports/coal-information-2019 (Accessed February 2019).

Electrification in South Africa, n.d.http://www.ee.co.za/article/department-energy-acknowledges-results-statssa-electrification.html. (Accessed in October 2019).

Energy Information Administration, (2018). eia.gov. [Online] (Accessed 2 November 2018).

HM Government, 2018. Digest of UK Energy Statistics (DUKES) [Online]. Available at: https://www.gov.uk/government/collections/digest-of-uk-energy-statistics-dukes. (Accessed 21 October 2018).

Institute of Energy Economics and Financial Analysis (2019). [Online]. Available at: http://ieefa.org/wood-mackenzie-met-coal-prices-likely-to-fall-in-2019-steam-coal-outlook-remains-challenging/. (Accessed in February 2019)

James, T., 2008. Energy Markets: Price Risk Management and Trading. John Wiley & Sones (Asia) Pte. Ltd, Singapore.

Natural Resources Canada, 2014. Energy Markets Fact Book 2014—2015, s.L. National resources Canada.

Natural Resources Canada, 2017. Oil Resources [Online]. Available at: www.nrcan.gc.ca/energy/oil-sands/18085. (Accessed 12 October 2018).

OPEC, 2018. OPEC Web Site [Online]. Available at: https://www.opec.org/opec_web/en/about_us/23.htm. (Accessed 28 October 2018).

United Nations, 2017. World Population Prospects: The 2017 Revision [Online]. Available at: https://population.un.org/wpp/Download/Standard/Population/. (Accessed 21 October 2018).

World Bank, 2019. Available at: https://papers.ssrn.com/sol3/papers.cfm?abstract_id=3442522.

3 ⣿

Crude oil

Crude oil is formed of complex combinations of hydrocarbon atoms (combinations of carbon and hydrogen atoms in long and short connected chains). These are the result of tiny sea plants and animals that died and became buried on ocean floors between 300 and 400 million years ago. Over time, these dead organisms became covered by layers of silt and sand deeper and deeper. The intense pressure and heat through time on this organic matter turned these deposits into the oil and gas embedded in source rock, typically sedimentary. The most common form of the source rock is a **black shale** that was created in ancient seabeds.

The temperature was the most critical component of oil formation, and the minimum temperature for oil formation is 65°C, which occurs at a depth of around 2100 m below the earth's surface. Oil deposits would normally be formed between 2100 and 5500 m depth; below that latter depth, and the higher temperatures found nearer the earth's core, would result in the formation of natural gas.

Oil and gas are buoyant and will float on water. So over time, the oil and gas deposits rise through sedimentary rock; this process is known as **migration**. This migration upwards continues until the oil reaches the impermeable rocks, which trap the oil and gas from rising any further. This rock formation is known as a **caprock**. The oil and gas rise to the top, forcing any water to the bottom of the reservoir formed; often, the water is saltwater. The most common forms of these oil traps are known as **domes** or **anticlines**, which result from a natural geological arch in the reservoir rock. These are shown in the following diagram — Fig. 3.1.

3.1 Exploration

Early oil finds were "discovered" through the seepage of oil to the earth's surface in various places, such as Pennsylvania, USA, Baku in Azerbaijan, and Trinidad. Seepages from the ocean seabeds led to oil being found off the coasts of Canada and Venezuela. At the start of the 20th Century, exploration techniques became more scientific with the use of surface topography and seismology.

Surface topography involves the mapping of sedimentary rock formations found on the earth's surface and then projecting this surface mapping into the subsurface rock. With seismology, sound waves are aimed from a source to bounce off subsurface geological surfaces, and the echoes from the formations beneath the earth's surface are recorded on the surface.

The seismic waves propagate efficiently and provide a high-resolution image of the subsurface, as the echoes or reflections travel back at different speeds depending upon the density of the rock layers, which the seismic waves encounter, as shown in Fig. 3.2.

Rural Electrification. https://doi.org/10.1016/B978-0-12-822403-8.00003-5

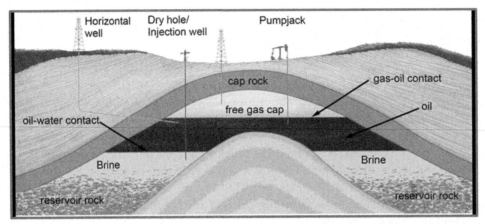

FIGURE 3.1 Anticline petroleum trap. *Source: Scientific Research (www.scirp.org). Used under Creative Commons Attribution License CC BY.*

The vibrations are recorded upon reflection by seismometers (microphones) on land, or hydrophones if exploring over water. The readings, in the form of graphical images, are then interpreted for evidence of oil or gas traps.

A potential oil trap is called a prospect, and once a prospect has been identified, the next step would be to drill a test hole into the top of the trap to determine if the prospect actually contains oil or gas. The cost of drilling is very high, so high losses will be incurred for every dry hole drilled, which concentrates petroleum geologists' efforts to ensure success.

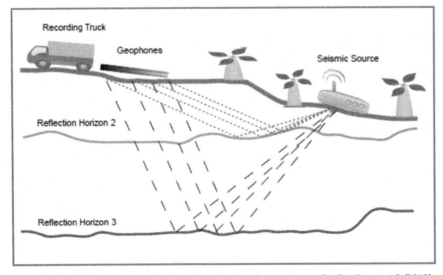

FIGURE 3.2 Seismic surveying. *Author's drawing based on an image by GeoExpert AG (2019).*

3.2 Production

If the geologist's predictions have been accurate, then an oil or gas strike will occur. A hole that drills into an oil or gas reservoir will be called a well. The oil will be under considerable pressure deep in the earth's crust, so once a well has been drilled, the oil will rise rapidly to the surface under considerable pressure. In the past, such oil first emerged as a gusher that sprayed oil all over the drill site, wasting valuable hydrocarbons. As the reservoir becomes depleted, the pressure remaining in the reservoir begins to reduce, and the flow slows down. Perhaps only half the reservoir's oil will be recovered under natural pressure. In order to ensure that the maximum volume of oil is recovered, another hole is drilled into the reservoir, adjacent to the well, and hot water or steam is pumped down, which pushes the oil out of the reservoir rock and up the well pipe. This technique is called ***enhanced recovery***.

Once the oil has been extracted from its reservoir, it must be safely transported to the refinery where it will be transformed into useful petroleum products such as gasoline, aviation fuel, and diesel. This primary transportation of crude oil can be done by pipeline, railcars, or large tankships. Transportation is discussed in greater detail in Section 3.7 of this Chapter.

One spectacular example of an oil field pipeline is the Trans-Alaskan Pipeline System (TAPS) that carries crude oil from Prudhoe Bay, in northern Alaska, some 1300 km south across Alaska's permafrost to Valdez on Prince William Sound. This was built in the mid-1970s after the 1973 oil crisis made the project economically justifiable. The 122 cm diameter pipeline enables crude oil to be loaded onto tankships in Prince William Sound for those longer periods of the year when Prudhoe Bay is icebound. The pipeline approach was considered less risky than building specialist icebreaker tankships to carry the crude oil.

3.3 Geography

The geography of oil is important to understand, as the largest oil reservoirs have been found some distances from the major consuming markets of Europe, North America, and the Far East. In the past, this has resulted in the major oil companies, predominately American, exploring for crude oil in the Middle East and then transporting their products to their own refineries located in their consumption markets. This, of course, simplified transportation of just one product (crude), but eventually led to the rise of nationalism in producer states that led to the sequestration of oil company assets as those states demanded a larger share of the wealth from their own natural resources. The following chart (Fig. 3.3) graphically demonstrates the different regions of the world for production and consumption.

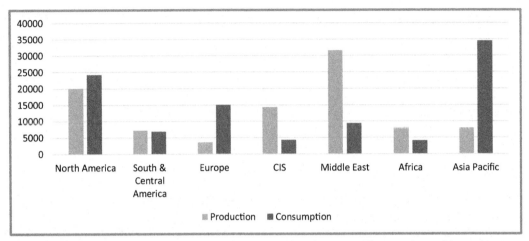

FIGURE 3.3 Global oil production & consumption by region (2017). *Data BP (BP, 2018); Graphic: Author Note: Units are 1000 barrels of oil per day.*

3.4 Geopolitics – OPEC; the rise of US shale production

The Organisation of Petroleum Exporting Countries (OPEC) was formed in 1960, with a mission to:

> *"co-ordinate and unify the petroleum policies of its Member Countries and ensure the stabilization of oil markets in order to secure an efficient, economic and regular supply of petroleum to consumer, a steady income to producers and a fair return on capital for those investing in the petroleum industry"*
>
> *(OPEC (2018))*

OPEC can use both pricing and production controls to influence the market and thence market prices. It first discovered this power during the Yom Kippur war of 1973, when crude oil export embargos were placed on the USA, the UK, and the Netherlands, with the result that crude oil spot prices quadrupled. OPEC started the setting of production quotas for its Members in 1983; these have not always been observed assiduously, and cheating on the quotas by overproducing appears to be a factor.

There are 15 member states – Algeria, Angola, Congo, Ecuador, Equatorial Guinea, Gabon, Iran, Iraq, Kuwait, Libya, Nigeria, Qatar, Saudi Arabia, United Arab Emirates, and Venezuela. Indonesia was a member from 1962 to 2009, and rejoined and also resigned during 2016.

The Shale oil and gas explosion since 2010 is beginning to transform the landscape of oil and gas production, not only geographically but politically. The predominance of US shale gas production has seen plans for LNG import terminals being quickly transformed into export facilities to create export opportunities for exporting potential. Internally, the US economy has benefitted greatly from much lower gas and energy prices, as low as one-third of their European equivalents. This forced Middle Eastern and African producers to have to find new market destinations for their oil and gas production.

Shale oil is already reducing US dependency on OPEC exports as the US works toward energy self-sufficiency. In late 2014, OPEC—largely at the behest of Saudi Arabia—tried to bankrupt the US shale oil producers by maintaining production levels at a time of static demand and a glut of supply. This resulted in crude prices falling from $115 a barrel in late 2014 to under $30 in the spring of 2015. The strategy has not worked: small shale oil producers merged or mothballed, to ride out the price storm. It is much easier to ramp up and down shale oil production than conventional fields, and now OPEC finds itself in a war of its own making. If they try to flood the market with overproduction, not only is OPEC's ability to do this in question, but shale producers have moved to a lower cost base and have become more resilient. Moreover, lowering oil prices does not sit well with many of OPEC's members who are overdependent upon petrodollars to fund their economy. Restricting OPEC supply will merely raise prices, and US shale oil Producers will simply pump more oil and rub their hands in glee.

3.5 Industry structure — IOCs, NOCs, GSEs

For a considerable period post-WWII, the oil industry was dominated, and some say controlled, by the so-called "Seven Sisters." This referred to the mainly American major oil companies: Esso Petroleum, Standard Oil of New York (Socony — later to be called Mobil), Standard Oil of California (Socal — later to be known as Chevron), Texaco, Royal Dutch Shell, and British Petroleum (BP). Together it was said these oil companies acted like a cartel and at one stage controlled an estimated 85% of the world's petroleum reserves. They were fully vertically integrated, owning and operating all the various stages of oil's value chain from reserves at wellheads, transportation, refining, and distribution assets.

Such companies are properly referred to as International Oil Companies (IOCs), and while their control of reserves was taken from them by the rise in nationalism in the oil producer states, they have become household brand names all over the world. During the late 1990s, there was a move to bracket the major oil companies as "supermajors.". At that time, and after some mega-mergers in the industry, these included ExxonMobil (a 1999 merger between Esso (Exxon) and Mobil), Chevron (which acquired Texaco in 2001), Shell, BP, Total Oil (of France), and US-based ConocoPhillips.

The IOCs are exchange-quoted, publicly owned enterprises, with ExxonMobil once being the single largest capitalized company in the world. They saw that there was no need to remain fully vertically integrated and have outsourced many of the intervening operations within oil's value chain, to concentrate on exploration and production activities, especially those located in inhospitable places or those requiring technical expertise and large capital investments. Recent estimates put the IOC share of proven reserves into a range of values between 6% and 10%.

After the formation of OPEC in 1960 and allied to a rise in nationalism as producer countries seized IOC assets, existing petroleum ministries in those producer countries began to be formed into state-owned National Oil Companies (NOCs). These NOCs are owned and operated in their entirety by their host governments. The precursor to this nationalistic fervor was Mexico's formation of Pemex in 1938, which took control of all foreign-owned oil assets in the country, and the example of the 1951-established National Iranian Oil Company.

The NOCs are operated in a much less transparent manner compared to the publicly quoted IOCs. Examples of NOCs include Saudi Aramco established the largest oil company globally (created between 1973 and 1980), Sonatrach of Algeria (created between 1963 and 1971), the Abu Dhabi National Oil Company (ADNOC), which was formed in 1971, Kuwait National Petroleum (1975), Petroleos de Venezuela (PDVSA), created in 1975, and the Nigerian National Petroleum Company (NNPC) of 1977. Predominately, these companies exist in the Middle East, and are believed to control over 90% of global oil and gas reserves and some 75% of production output.

A third category of classification has been created, known as Government Sponsored Enterprises. This refers to NOCs where the host Government has sought to raise additional capital from public subscription, and thus sold-off a proportion (usually a minority share) to achieve this aim. This has created oil companies that are part-Government and part-public owned. This can cause a dichotomy in business aims, as the public shareholding would be expecting profits to pay dividends, whereas the Government will still expect the GSE to serve its national interest, a potential conflict of goals and objectives. Examples of GSEs are Petroleo Brasileiro (Petrobras) of Brazil, OMV Petro of Romania, Statoil of Norway, and Aramco of Saudi Arabia.

Some GSEs have spurned IOCs as they grew and started to acquire assets and operations outside of their host country; many of these emanated from Russia in the aftermath of the break-up of state control, such as Gazprom, Rosneft, and Surgutneftegaz, but also PetroCanada (which was a state-owned enterprise until its 1990 take-over by Suncor of Alberta, Canada), and the Oil and Natural Gas Corporation of India (ONGC).

The oil industry also possesses a host of independent producers who concentrate mainly on exploration and production and are based in the US. These organizations tend to concentrate in on-land and near-offshore locations. They include Anadarko Petroleum (US), BHP Billiton (Australia), Chesapeake Energy (US), Maersk (Denmark), and Noble Energy (US). Oilfield service companies, such as Schlumberger, Halliburton, Transocean, Baker Hughes (all US), TransCanada, and Saipem of Italy, exist to supply specialist skills and technology to assist halting the decline in output from mature fields and build and operate drilling rigs. They exploited an opening in these aspects as IOCs either exited countries or began their outsourcing programs. They often work with NOCs, as the service companies have no interest in crude oil acquisition, and thus, producer governments do not see them as a threat or having conflicting interests as they would with an IOC partner.

3.6 Fiscal systems: R/T; PSAs/PSCs; service agreements

"Governments take as much money as possible from one part of the citizens to give to the other"

Voltaire (French Philosopher) 1764

Typically, the mineral rights for oil and gas deposits found underground or offshore are in the ownership of the host government or state, from whom access rights must be sought before legal access is consented to. The major exception to this general rule is in the USA and Canada, where the landowner also owns the mineral rights to any deposits beneath the surface of the land. The mineral rights of a country are usually enshrined in the Constitution of the country and may be administered by either a Department of Energy (or Mines and Minerals) or the National Oil Company.

The first type of fiscal system that governed the monetary aspects of oil exploration and production was ***concession*** systems, also known as ***Royalty/Taxation*** systems. In return for an exclusive access lease fee, the lessee took the title to any oil produced at the wellhead and was free of constraint from the host Government as to how much oil was produced or where it sold that oil. The producer was liable for a royalty upon commercial production and paid income tax on its profits. Sometimes, the royalty rate was set at a fixed rate per tonne, and as oil prices increased, the host Government was automatically excluded from the upside benefit of rising prices. Concessions were granted for very long periods of time; for instance, in 1939, five oil companies were granted exclusive development rights over the whole of Abu Dhabi for 75 years.

Royalty/Taxation systems still account for around half of the fiscal systems in use, but are now for much shorter periods, have stricter terms that offer much better protection of the host Government's interests, and require specific exploration activity from the oil company. A survey of typical fiscal system elements showed that Royalty rates under concession agreements average out at just under 10%.

After the rise in nationalism seen in oil producer states, and the formation of OPEC in 1960, a new fiscal system was introduced to redistribute the control, ownership, and profits from oil exploration and production away from IOCs and in favor of the host Governments. This system was known as a ***Production Sharing Agreement (PSA)*** (sometimes Production Sharing Contract (PSC)), a form of contractual system. Under the PSA, the major change from the concession system is that the host Government maintains ownership of the oil produced far beyond the wellhead. The IOC bears all the risk of exploration and production, but part of the oil produced is allocated to cover the costs of development and production (cost oil). The residual oil (or profit oil) is split between the host Government and the IOC on a pre-agreed percentage split; typically, this can be 60%/40% in the host Government's favor. The producer is allowed to lift his share of the profit oil and all of the cost oil as his entitlement. Costs that can be charged to the fiscal system are usually capped or limited to a fixed percentage of the gross revenue, as host Governments seek to protect themselves from creative or fraudulent accounting by the IOC.

PSAs have proved very profitable to host Governments, when oil deposits are already known and exploited; for instance, Venezuela's government "take"[1] free of cost oil is between 88% and 93%. The biggest difference can be said to be in the host Government's guaranteed share of revenues (also called the Expected Royalty Rate), which is the minimum share of revenue that a government might expect in any given accounting period, on the assumption that cost oil is at the maximum level permitted and that the IOC (legally) pays no income tax in this period. Under the R/T system, the typically guaranteed share of revenue would be 10%; under PSAs, this is tripled to 30%.

Exploration and production can also be arranged under **risk service contracts**, whereby the IOC provides some or all of the capital expenditure required and is paid like any contractor or service provider. Similar systems, whereby the IOC receives a flat fee for development services, would be called a **service agreement**. In both cases, the ownership of any crude produced remains the property of the host Government.

3.6.1   The global market for oil

Most of the developed industrialized world would not exist today in the same form if it were not for oil's contribution, especially to transportation and the wealth it has created in the past through international oil companies and in the present for the producer country economies. Part of this interdependence between oil and economic development has been made possible by the easy transportation and storage of crude oil and the refined petroleum products, the majority of which are in the liquid state at ambient temperatures and can be moved around the globe very easily.

In the early years of the oil industry's development, this was mainly the shipment of crude from the Middle Eastern oilfields to refineries operated by the international companies, situated in their homeland markets of North America and Europe. In turn, and in some way due to the fungible nature of the basic petroleum products, global trade has developed, and while it would be true to note that the transportation of both crude oil and petroleum products is kept to a minimum, as transportation always costs money, there is an interconnected pricing relationship between all the regional markets. This is unlike natural gas, where the ability to liquefy the gas for the ease of transportation has only relatively recently been perfected. Whenever the oil market is discussed, both producers and consumers, and all the middle-men, wish to obtain a "fair" price for their oil, whether selling or buying. A fair price for a seller may not, of course, be interpreted as a fair price paid by the buyer of that seller's oil. For refiners, a fair price would be that the cost of crude oil that the refiner can profitability transform into petroleum products. Consumers might define a fair price as that, which can sustain current levels of usage without affecting their current quality of life. A producer of oil would expect a fair price definition to include the costs of exploration, the costs of production, and the transportation costs of that crude oil to its destination market. In addition, the "fair" process must be sustainable over the long-term and secure adequate funding and foreign income

---

[1]"take" = Royalty plus Government share of Profit Oil plus Income Tax paid by IOC.

for their national budgets. While these aspirations are not always achieved, the price paid for oil will be reflecting the basic supply and demand fundamentals. These are defined in the next section.

3.6.2   Supply and demand fundamentals & drivers

In the short term, crude oil and the vast majority of petroleum products display the classical price inelasticity of any commodity without a ready substitute. In other words, in the short-term, demand will not be affected by the level of price or price movement. In the longer-term, however, there are specific supply and demand drivers for oil; these are detailed below in Box 3.1.

3.6.3   Characteristics of crude oil

Crude oil, as a product in its own right, is a fairly useless commodity but derives its value from the range of petroleum products it can produce when it is refined and transformed. It would be fair to say that the price of crude is now driven by the price of products, in turn, influenced by the long-term supply and demand fundamentals of those products. Crude oil is a complex mixture of hydrocarbons, and the composition of that mixture changes from oilfield to oilfield, and this quality differential can mean that some crudes are more highly prized than others, as will be discussed in the next few sections.

The US Energy Information Administration defines crude oil as:

"A mixture of hydrocarbons that exists in liquid phase in natural underground reservoirs and remains liquid at atmospheric pressure after passing through surface separating facilities. Depending upon the characteristics of the crude stream, it may also include:

1. *Small amounts of hydrocarbons that exist in the gaseous phase in natural underground reservoirs but are liquid at atmospheric pressure after being recovered from oil well (casing head) gas in lease separators and are subsequently comingled with*

BOX 3.1 Insight: the fundamentals of oil supply and demand

SUPPLY: Quantum of proven oil reserves and whether these are being depleted at a faster rate than that of new discoveries. Actual production levels, especially in relation to proven reserves. Production quotes set and endorsed by OPEC and their allies. Levels of exploration and production capital expenditure authorized by oil companies. Improvements in yield from existing fields will increase supply. Disruptions, especially those that occur unexpectedly—these can be both natural (weather-related) or political—both would have a detrimental effect on the supply of oil.

DEMAND: Obviously, a function of consumption is, in turn, heavily influenced by economic growth, increasing industrialization of developing countries; all these, in turn, are positively influenced by population growth. In the intermediate-term, a lack of enough refining and storage capacity would limit the demand growth potential of petroleum products. Negative fundamentals for demand will include the increasing greater fuel efficiency of motor engines, and governmental legislation that reduces reliance on fossil fuels and promotes renewable and alternative energy in a move toward a low carbon society and to meet climate change targets.

the crude stream without being separately measured. Lease condensate recovered as a liquid from natural gas wells in lease or field separation facilities and later mixed into the crude stream is also included

2. *Small amounts of nonhydrocarbons produced with the oil, such as sulfur and various metals*
3. *Drip gases, and liquid hydrocarbons produced from tar sands, oil sands, gilsonite, and oil shale"*

(Energy Information Administration, (2018))

All crude oils have a different make-up with differing proportions of short- and long-chain hydrocarbons, from single carbon molecules through to very long chains containing more than 70 linked carbon molecules. In addition, different crudes have varying levels of impurities, sulfur, nitrogen, and metal compounds. There are almost as many crude oil types as there are oil fields.

3.6.4   How crude oil obtains its value

Crude oil, as stated before, has no value in its natural form, but its value derives from the products it can help create and the proportion of the higher value products (typically the transportation fuels) that an individual crude oil can yield. The higher the quality of the crude relative to the benchmark crudes that are used as price reference points, then the higher the price premium that particular crude may command.

In theory, refiners have a simple choice between a heavy and sour crude that can be acquired at a significant discount to the market price for WTI or Brent, but this will cost more energy to process and will require a higher capital expenditure investment in the refinery build and configuration.

The value of crude oil will fluctuate with demand and supply factors, as well as quality issues and differentials. In general, the differential between light sweet crude (higher quality) and heavy sour crude (lower quality) will widen with increasing global oil prices. Typically, the differential range is between 15% and 25% of the price of light sweet crude.

3.6.5   Crude oil quality — sweet/sour; heavy/light

There are two primary measures of crude oil quality, which will influence the price paid to producers by the market — largely determined by refiners with configurations that can either run that crude without blending, or whether the quality is such that blending with (usually) a lighter grade is necessitated.

These two measures are known as the sulfur content and API Gravity (see Box 3.2).

Crude oil with a sulfur content in excess of 0.5% by weight is known as a ***sour*** crude, whereas sulfur content less than 0.5% qualifies the crude oil to be termed ***sweet***.

BOX 3.2 API gravity

American Petroleum Institute measure of specific gravity of crude oil or condensate in degrees. An arbitrary scale expressing the gravity or density of liquid petroleum products. The measuring scale is calibrated in terms of degrees API; it is calculated as follows.

Degrees API = $(141.5/sp.gr.60 F/60 F) - 131.5$.

If the API gravity of crude oil is higher than 31.1° API, then it is known as *light*; an API gravity below 22.3° API means the crude oil is termed *heavy*. The middle ground is naturally *medium*, but this description is not as commonly heard as the other two terms.

Some of the better-known crudes are shown in the following Table (Table 3.1) to show these primary characteristics in each case.

3.6.6 Importance of quality and correlation to the price obtained

In addition to the geographical location, the quality of the crude oil will define the yield and quality of the products produced in the refining transformation process. The lower the quality and the greater the proportion of impurities will mean that the costs of refining, cracking or reforming, and tertiary treatments in the refinery will be higher. Therefore, a heavy sour crude oil will be less attractive to a refiner than a light sweet crude and will, therefore, have to be sold at a discount to the higher quality crude.

High-value crude oil will invariably be light and sweet, from which it is easier (and less costly) to obtain a greater yield of gasoline. Light crude oil will perhaps naturally[2] yield 45% of its volume to gasoline, whereas a heavy crude might only produce as little as 15%–20%. Most of the European refineries built in the 1950 and 1960s have been constructed and configured to refine light sweet crudes such as North Sea Brent Blend and Nigerian Bonny Light. US refineries are usually configured to maximize the gasoline yield.

Table 3.1 Primary characteristics of well-known crude oils.

Crude oil	API gravity	Sulfur content	Classification
West Texas intermediate	37–42	0.42%	Light, sweet
Brent	38	0.40%	Light, sweet
Dubai	31	2.00%	Light, sour
Forcados (Nigeria)	29.5	0.20%	Medium, sour
Minas (Sumatran light)	35	0.08%	Light, sweet
Tapis (Malaysia)	45.2	0.0343%	Light, sweet
Isthmus 34 light (Mexico)	33.74	1.45%	Light, sour
Marlim (Brazil)	20.1	0.70%	Heavy, sweet
Maya heavy (Mexico)	21.3	3.40%	Heavy, sour

[2]"naturally" = in this context, through normal distillation, i.e., without secondary cracking or reforming.

New-build refineries in the Middle East, India and the Far East have more recently been built with a wider capability to the "*slate*" of crudes they can run; this is future-proofing as most new finds of conventional crude will be heavier and sourer than Brent or WTI.

Evidence shows that generally, API density is trending downwards and sulfur content upwards. This is due to new large oilfield production from Canada (heavy and sour from tar sands), from South America (also heavy and sour) and the Middle East (mainly medium and sour). Historically the light and sweet crude produced in Europe and Asia more than balanced this out globally, but as production declines in these two areas, refiners having to accept that the days of abundant light and sweet crude oil availability are coming to a close.

3.6.7   Benchmark crudes

All crude oils have different physical properties and characteristics and are located all over the globe. It would be impractical to try and establish a globally traded market for crude if every grade were traded individually, so the industry has evolved in such a way that a only few major crude oils are used to price much of the world's output. These crudes are known as **benchmark crudes**, effectively price markers for other crudes, which can be traded at a premium or discount to the benchmark, depending upon the quality of crude and distance from the destination market.

The three most widely traded benchmarks are West Texas Intermediate (WTI), North Sea Brent Blend and Dubai Light. The trade benchmarks in the first two far outstrip the importance of the Dubai benchmark.

WTI yields a high proportion of gasoline, an important quality for North American refiners, and is a sweet crude that is used to benchmark all crude imports, irrespective of origin, to the USA. Production of WTI is in decline with a refining concentration in the American Midwest and the Gulf Coast.

Brent Blend has derived its name from the Shell Brent field in the UK's North Sea and has been traded since 1988.

Brent Bend references a combination of some 15 crude oils from the Brent and Ninian fields in the North Sea, and is a light sweet crude, like WTI, but not quite as light and sweet. Just as in the case of WTI, Brent production is in decline, but it is said to be used as the benchmark for two-thirds of the global crude output.

OPEC has tried to establish a rival benchmark crude – the OPEC Reference Basket. This is a weighted average from the crudes of its producer members' and informs OPEC's production output, which has the objective of keeping the reference basket price within certain, agreed limits. Other attempts to establish a Middle Eastern benchmark failed because there was no underpinning of physical crude oil deliveries, and pure trade speculation will not ensure the stability that a benchmark crude requires.

Despite being a major trading and consuming area, Asia has not established a specific futures exchange-traded grade or any benchmark crude oil for the region.

The market price relationship between WTI, Brent Blend, and Dubai Light tends to mirror each other over time, but typically WTI will trade at a $2 premium to Brent and a $6 premium to Dubai Light and the OPEC Reference Basket. However, short-term supply and demand conditions can cause the markets to trade WTI at a lower price than Brent.

This trading differential is often influenced by the availability of refining capacity and the ability for any given refiner to process heavier and sour crudes. As refiners have been unable to justify the capital expenditure to upgrade their refinery capability, light sweet crudes that generate a higher yield of gasoline and distillates when refined—such as WTI and Brent—are increasingly sought as the optimal crude oil of choice.

3.6.8   Trading centers – ICE, NYMEX, Singapore

Oil and gas trades are dominated by three major markets in London, New York, and Singapore. These are well-established centers where many active buyers and sellers give rise to highly liquid markets. Such markets provide the opportunity for leveraged derivative investments, such as oil futures, and attract large inward risk capital. These markets provide low cost of entry and exit, and with exchange house regulation and guaranteeing the performance of the futures contracts traded on those exchanges.

New York's Mercantile Exchange (NYMEX) was formed as far back as 1872, initially trading in butter and cheese, and is currently owned and operated by the CME Group of Chicago.

The London trading center started in the 1980s with the International Petroleum Exchange trading initially in gasoil futures and adding Brent crude oil futures trading in 1988. It was acquired in 2001 by the American-based Intercontinental Exchange and is commonly now referred to as ICE London.

SGX Singapore was formed in December 1999 by the transfer of assets owned by former exchange companies - Stock Exchange of Singapore (SES), Singapore International Monetary Exchange (Simex), and Securities Clearing and Computer Services Pte Ltd. (SCCS).

3.6.9   Price discovery

Price discovery refers to the means by which a commodity, in a freely traded market, is valued, such that when the classical economic theory of supply and demand is applied, the price is a true representation of that commodity market. In other words, if demand exceeds supply, then the price of the commodity would be expected to rise until such time as new sellers enter the market to trade at the higher price, and the supply and demand curves meet and intersect. In economic theory terms, this price level is known as the price at which the market clears, as sellers match buyers. The reverse obviously applies; commodity prices must fall when supply exceeds demand.

Good price discovery is assured when that commodity market is highly liquid, with many active buyers and sellers participating in that market all the time. Where there are just a few participants in a market, then manipulation can occur if a trader withholds from buying or selling to create a false market position to affect the price. However, honest speculators do play an important part in improving market liquidity, and thus aiding better price discovery.

In the oil futures markets, the derivative contracts fulfill some important requirements to aid greater price discovery and transparency. Sellers of oil are able to view the forward market and get a reasonable impression of future prices for that commodity, and thus, make a more informed decision about whether to sell their oil today, or place into storage and sell in the future, or to enter into a futures contract to sell the oil at a later date. Consumers of oil and oil products are also able to understand the forward market, and manage any price-risk and take out hedging contracts on the derivative markets.

As futures markets dampen the peaks and troughs of supply and demand, volatility in the commodity is also reduced. This aids long-term planning, and in the oil industry where infrastructure and transformation projects require long time cycles to implement and for amortization, the smoothing of short-term volatility assures the risk-takers that the downside risk will tend to be short-lived.

As futures contracts approach their expiry, or settlement, date, the price at which they are traded also trends toward the spot, or cash, market price for that commodity. Upon expiry, it would be an assumption that the value of a futures contract that has to be performed immediately (to take or deliver "*wet*" or physical barrels) would be expected to have the same intrinsic value as a physical contract for that same commodity. If this principle was not prevalent, then an artificial arbitrage in that commodity would be created, whereby a margin could be present between the paper and the wet barrel contracts, and by entering into back-to-back contracts, a profit could be immediately generated for no real benefit to the industry producers or consumers.

The major benchmark crude markets and the major petroleum products in which NYMEX and ICE futures contracts are available do satisfy all the criteria required for price discovery. In both markets, there are a large number of players; the financial leverage available to become involved in the futures contract markets allows organizations outside the oil industry to participate; a larger number of players increase liquidity; the futures markets do reflect the value of the underlying physical commodity, and the market (in its widest sense to include the traders and the players) has evinced a sense of confidence over many years of successful trading.

The linkage between the futures and the spot market prices for the same commodity is now well established, and the relationship between the two prices will exhibit both parallelism and convergence. Parallelism defines the high correlation between spot and futures prices, with the same demand and supply factors impacting on both spot and futures contract prices. Convergence is the terms used to describe the already-discussed meeting of spot and futures contract prices, as the futures contract approaches its expiry date.

On-exchange traded and regulated markets will provide good price transparency for a commodity such as oil or gas. When that is combined with a very liquid market, in which the sheer level of traders are then reported widely and openly, it delivers price discovery.

In the case of the NYMEX and ICE oil and gas markets, the confidence in the futures market prices as a representation of future expected physical commodity prices is now assured. Price discovery is an essential prerequisite for this preeminence that has been established over many decades; it will be a continuing requirement for the efficient and effective operation of oil commodity futures markets.

3.6.10 Contango and backwardation

Two traditional market-condition terms are introduced at this point, ahead of the discussion of derivative and futures prices. If a commodity has both an immediate physical market price—sometimes referred to as a **spot** price—and prices established for future delivery of the same commodity, then a graphical representation can be drawn of the forward market of the futures commodity against the spot price.

The alternative to buying a physical spot commodity and immediately delivering that commodity is to purchase and hold it in storage as physical inventory for physical delivery at some future point in time. This action costs money, of course; there will be the cost of storage, the interest payable on the cost of inventory, and an opportunity cost associated with this strategy. These costs are known as the **costs of carry**. When the forward futures prices are above the physical spot price plus this cost of carry, then the market is said to be in **contango** (see Fig. 3.4). The incidence of a contango demonstrates that the commodity traders expect future supplies of that commodity to be scarcer than current availability.

A formal definition of contango could be: "*A percentage paid by a buyer of stock to postpone transfer to a future settling day*" (Oxford Living Dictionaries, 2018). In other words, the buyer, who wishes to settle the commodity price today, but does not want, or cannot, accept physical delivery until [say] 3 months hence, would pay a premium to the seller of the community to delay physical delivery.

The exact opposite market conditions—when forward futures prices are lower than the physical spot market—is known as a market in **backwardation** (Fig. 3.5). A backwardated market occurs when there is a current supply shortage of the commodity, which is expected to be resolved in the future. The current shortage drives up spot prices as there will be more buyers than sellers, but as the circumstances, which caused the supply shortage, begin to be corrected and supplies increase, the balance will swing toward the sellers who will be forced to lower prices to find buying trade.

A formal definition of backwardation could be:

> "*the seller's postponement of delivery of stock … with the consent of the buyer upon payment of a premium to the latter*"
>
> (Merriam-Webster, 2018)

In other words, the seller of the commodity effectively pays a fee to the buyer - in the form of a discount against the spot price – to defer physical delivery until a later date.

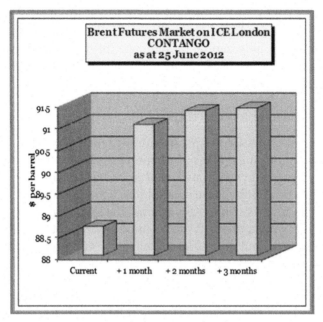

FIGURE 3.4 Contango market.

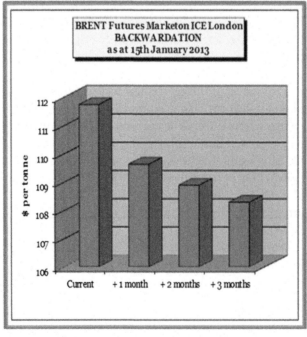

FIGURE 3.5 Backwardated (or inverted) market.

Generally, the later the delivery, the larger the premium or the lower the price. In this way, an illiquid market in the very short term does not stop a seller from trading; by offering a discounted future price, he can gain a sale by deferring physical delivery until commodity supplies have overcome the current shortage. The higher spot price reflects the asset of holding physical inventory at a time of a shortage in the market; this can sometimes be referred to as the **convenience yield**.

The traditional and historical oil market condition is for a contango market in the short-term, with a backwardated market for the longer forward future delivery dates for that commodity.

3.6.11 Derivatives and futures

Derivative contracts are so-called because they **derive** their value from the value of the underlying commodity that is traded in physical markets. These physical trades are sometimes referred to as "**wet**" barrels to distinguish them from the derivative and futures markets, which are paper trades only, and thus known as "**paper**" barrels.

The appeal of the derivative market is that one can participate in oil trade without ever having to make or take delivery of physical oil. It is important to note that when holding a WTI futures contract to expiry, the contract holder must take or make delivery of physical oil barrels to Cushing, Oklahoma, USA; all Brent futures are financially settled and never go to physical delivery. Trading in paper barrels usually never conveys ownership of the actual oil barrels underpinning the derivative contract. While less than 5% of all paper trades are never held to expiry, and thus, converted into physical oil, the linkage between the paper and wet barrels determines that the derivative market plays an important part in providing liquidity and price discovery and transparency.

Derivative contracts can be traded via on-exchange trades, where the exchange house acts as an intermediary and guarantees the trades on both sides of the contract by only authorizing brokers of sufficient standing and financial perspicacity to conduct trades. There is also the ability to conduct what are known as Over-The-Counter (OTC) trades; as there is no middle party to guarantee performance, these carry a great additional risk for both parties, and are therefore, mainly conducted between oil companies and traders.

The three major derivative contracts are futures, swaps, and options.

An on-exchange trade **Futures** contract, the buyer contracts to purchase a standardized volume of oil at a given date in the future, and the contract price is settled when the contract is entered into. A party that buys a futures contract is said to be "*going long*"; the party that sells is said to be "*going short*".

Buying a futures contract implies that the buyer believes the price of the commodity bought will increase in the future; effectively, the futures contract fixes the price today as a hedge against prices rising during the period during which the futures contract can be traded or liquidated.

On-exchange futures contracts are inflexible as the terms are imposed by the exchange house – but such rigidity means that many contracts of exactly the same size and format are traded every day and thus provide easy comparisons and aid price discovery and transparency.

A crude oil futures contract in either WTI or Brent Blend, as traded on the New York Mercantile Exchange (NYMEX) or London's Intercontinental Exchange (ICE), is for one lot, representing 1000 barrels of crude. Both NYMEX and ICE are governmental-regulated, as well as being publicly quoted companies. The CRB Commodity Yearbook (2008) evaluated that the NYMEX and ICE annual contracted volumes for crude oil and the major petroleum products accounted for around 98% of the global trade.

The liquid futures contracts that are traded with very high liquidity on both NYMEX and the ICE are WTI Crude Oil, Brent Blend Crude Oil, and Heating Oil; additionally, NYMEX offers a futures contract on Gasoline.

For initiating a trading position in a futures contract, your broker will be required to post an "*initial margin*", the size of which is determined by the exchange house; this could be at a level that offers a 10:1 leverage ratio, in other words, you could enter a futures contract with the full value of $100,000 for as little as $10,000.

In order to regulate the brokers and their trading exposure, a daily "*mark-to-market*" process values all open trades against the settlement prices for the futures contracts. This results in a variation margin when the market has moved against the futures position. If this is an adverse movement and has gone against the trader's (and thus their client's) position, then a loss will have been suffered (on paper, at least) and a "*margin call*" will be made by the exchange house to ensure that day's market movements have been covered.

Futures positions are often closed out by entering an equal and offsetting contract and notifying the exchange house, which then uses the offsetting contract to cancel the original position taken. If the volumes on each contract match, then the contract is said to be "*squared out.*" An inexact match would suggest some speculation on the part of the client.

A contract for future trade between two organizations directly—Over-The-Counter—would be a *swap* agreement, which is similar to a futures contract. Such swap contracts provide for infinite flexibility, against the on-exchange futures contracts, which are heavily standardized and transparently traded. Some standardization of contract terms and conditions can be applied, using contracts and clauses provided by the International Swaps and Derivatives Association (ISDA).

Swap contracts receive their name from the fact the transaction is swapping oil price movements between the two counterparties to the contract. In its simplest form, this will involve one commodity or product and swapping a fixed price (known at the time of entering the swap) for a future price to be determined during the agreed pricing reference period. A counterparty that is said to be going long in a swap, will be expecting prices to rise in the future; the swap contract, therefore, will increase in value as the commodity price rises, and will lose value as price falls away. To arrive at this position, the counterparty will be buying the fixed leg of the swap and selling the floating leg. Consumers of oil, such as airlines, would be the likely buyers of a swap, as they seek to protect themselves from the risk of future prices in their required commodity rising in the future.

Conversely, producers of oil, such as either exploration and production companies or refiners, seek protection against oil prices falling in the future. A swap contract that gains in value when prices fall meets their requirements, and thus they become sellers of swaps. Such counterparties will, therefore, sell the fixed leg of the swap and buy the floating, known as going short.

OTC swap contracts are normally settled in cash upon expiry, usually against an average price over a month or quarter as negotiated at the outset, when the floating price reference period is agreed between the counterparties. The daily prices used to determine such settlement are those published by exchange houses or trade journals for benchmark crudes or the traded petroleum products.

The theory shown above details the mechanics of the most simple swap contract — this is often known as a "***vanilla***" Swap; other forms of swap contracts exist, such as a "***differential***" or "***spread***" swap, where a fixed price for one product is swapped for the floating price in another product; one can also arrange a swap contract for the same product but over different time periods. In short, swaps are highly flexible derivative contracts that can be tailored to the requirements of the counterparties, and thus, overcome the inflexibility of on-exchange futures contracts.

The final derivative considered is the ***option*** contract. The holder of an option contract has the right, but not the obligation, to buy or sell the commodity at the agreed price during the agreed period when the option can be exercised. To gain this benefit, the option holder will have paid a one-off premium to the contract counterparty. An option contract can be viewed as a sort of insurance policy against which you can claim if the markets move against your position. Just like insurance, once you have paid the premium, but do not claim, the premium remains paid. The buyer (or holder) of the option is only exposed to losing the premium paid, whereas the seller could have unlimited losses unless they have hedged their position to offset these potential losses. The price agreed at the outset for exercising the option is called the "***strike price***".

There are two basic types of options contracts: a "***put***" and a "***call***"; in the former the underlying value of the option will increase when prices fall, and in the latter when prices rise. The holder of the option will only receive a pay-out if they exercise their option — this would normally be given when market prices are above the strike price for a call option (Fig. 3.6), and below the strike price for a put option (Fig. 3.7).

Consumers of oil, such as airlines and shipping companies, would be most interested in call options, whereas a put contract will have appeal for an oil producer or refiner.

Payoff diagrams are useful for a greater understanding of how options contracts work. In the examples shown for both a call and a put swap contract (Figs. 3.6 and 3.7 respectively), the strike price agreed between the two counterparties is $76.00 per barrel.

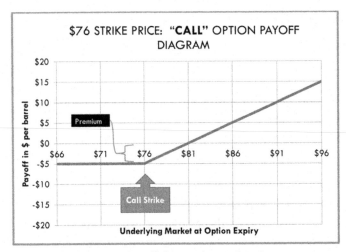

FIGURE 3.6 Payoff diagram for a call option.

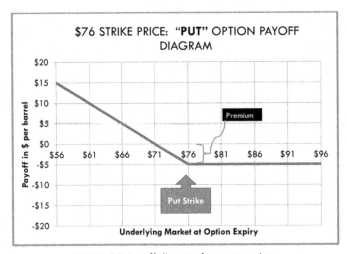

FIGURE 3.7 Payoff diagram for a put option.

3.7 Transportation

The oil and gas supply chain involves all the transportation models associated with the movement of liquid and gaseous commodities from source to end-user location. From the earliest days of wooden pipelines in the USA to modern-day specialist tank-ships capable of transporting highly cooled natural gas as Liquefied Petroleum Gas halfway around the world from gas fields in the Far East and Australia to the consuming markets of Europe and Northern America, the industry's ingenuity of moving its core oil, gas, and products in the most optimal, cost-effective manner has ensured that consumers can buy and utilize useful petroleum products, without pause as to how those vital fuels have made their way from oil and gas fields through the refining processes to their automobiles, offices, factories, and their homes.

In this section, we will examine the significant transportation models in turn, then discuss an often-overlooked part of the oil supply chain, terminalling, and finally discuss the economics of each of the transportation modals, relative to each other.

In the clear majority of cases, petroleum products usually arrive at their final destination by road tanker, even if the upstream chain could have featured every single model available. As the supply chain can account for up to 70% of the overall costs of the product at the end-user, optimization of transportation has occupied the management of oil companies for many years.

What is seen nowadays in the developed nations is a highly sophisticated and integrated system, operating in a seamless manner; this is despite the individual stages in the logistical and transportation stage now tend to be outsourced to specialized third-party providers to provide greater critical mass and technical specialism and expertise.

In short, transportation is a highly specialized operation in the oil and gas industry that requires advanced management and a technical skillset.

3.7.1 Crude oil transportation (oil field gathering systems; pipeline; crude oil tankers)

Although the streets of Baghdad were paved with tar[3], probably from crude oil that had bubbled up to the surface, it is an invariable fact that most of the large crude oil deposits—for instance in the Middle East and Venezuela—are some thousands of miles from where the consuming markets of the Far East, North America, and Europe are situated. It is inevitable that either the crude oil had to be shipped to refineries in the consuming countries or that refineries were built near oil fields, and the refined petroleum products transported great distances. An axiom of the transportation world is that every time you handle a product, it costs money; so, transporting just one product – crude oil – between the Middle East and the USA is very much more optimal, and costs are lower than transporting smaller batches of the myriad of refined products the same distance.

In a typical oil field, there may be many hundreds of individual oil wells producing crude. Production from these wells is transported in what is called a ***gathering system*** to a collection point for onward transmission to a refinery or a tanker load point. These gathering system lines are small, typically 2″ (50 mm) to 8″ (200 mm) in diameter, whereas the trunk or transmission crude pipelines are to be found with diameters mainly ranging from 8″ (200 mm) to 24″ (600 mm) diameters.

From the oilfield, crude oil is transported to the refinery by tankship, so a load point must be constructed to facilitate the safe transfer of the crude oil in volumes of anything between 60,000 and 500,000 tonnes. These load points could be onshore facilities, but more often, considering the size of supertankers, offshore platforms, or single-point

---

[3]Earlier than the 9th Century; also, Marco Polo, the explorer, commentated about oil well production in the region in the 13th Century.

moorings must be used, as a depth of around 75 feet of water is required for a fully laden Very Large Crude Carrier (VLCC). Offshore facilities can be 10 to 20 miles away from land and use a crude pipeline to connect from the oilfield holding system. Shore-based pumps usually load the tankship from the pipeline.

Over the years, the evolution of crude-carrying tankships has seen an evolution to bigger and bigger carrying capacities, which in the shipping world are noted in deadweight[4] tonnes. The following Table denotes the names, categories, and capacities of crude oil tankships (Table 3.2).

Only a few ULCCs remain in service because of their huge size and draft requirements and are found today, transporting crude from the Persian Gulf to Europe and from America to Asia.

The economic pressure to build adequate loading and offloading facilities is enormous. A VLCC loading in the US Gulf requires three to four *Aframax* sized tankships to lighter the oil from shore to the VLCC − this can cost around $250,000 per *Aframax* tanker, so finding solutions to save this additional $1 million cost per voyage is considerable and there are over six major loading infrastructure projects due for completion in the next four years in Texas, USA, alone.

Operating costs for tankships are usually referred by the description of "***chartering cost***," which is the act of hiring a ship to carry a cargo. There are a variety of chartering agreements, within which the key concept of freight rate arises; this is the price for the carriage of a specific cargo. This rate could be a lump sum figure, by a time-charter rate, where the cost is agreed on a per *diem* basis, a rate per tonne of cargo, or by reference to Worldscale rates. These Worldscale rates are annual published rates used as a benchmark by the shipping industry to agree with spot market charters. Worldscale benchmarks establish a benchmark for carrying a tonne of product between any two ports in the world. This can be used as a reference point between two charter parties when negotiating a specific voyage rate.

Table 3.2 Tankship classifications.

Tanker class	Deadweight tonnes	Capacity in bbls	Indicative new build cost
Panamax	60,000–80,000	500,000	$ 46 m
Aframax	80,000–120,000	750,000	$ 61 m
Suezmax	120,000–200,000	1 million	$ 61 m
Very Large Crude Carrier (VLCC)	200,000–320,000	2 million	$ 120 m
Ultra Large Crude Carrier (ULCC)	320,000–500,000	Up to 4 million	

Notes: *Panamax*, denotes the largest loaded tankship that can pass through the (old) Panama Canal; *Aframax*, so-called, derives its name from the Average Freight Rate Assessment tanker system established by Shell in the 1950s, and forms the standard for contract terms - it has no geographical significance; *Suezmax*, denotes the largest size of tanker that can pass through the Suez Canal when fully laden.

[4]Deadweight tonnage is the displacement of a fully-loaded ship, minus the weight of the ship itself. Cargo carrying capacity of ship is c. 95% of DWT; the balance of 5% representing fuel, water, ballast, crew, and provisions.

Pipelines are constructed not only to carry crude oil from field to ship and from offloading points to refineries but also petroleum products from refineries to inland distribution hubs or terminals. Construction costs can vary with length of pipeline, and less obviously, the terrain traversed. They can be inobtrusive if buried underground, but in certain locales — the Trans-Alaska system, for instance — may run above ground for ease of maintenance and integrity inspection. The Baku-Tbilisi-Ceyhan pipeline, constructed to carry crude oil from the Azerbaijan and Georgian fields to the Ceyhan, Turkey load point on the Mediterranean Sea, is 1100 miles (1800 km) in length, totally underground, and can carry one million barrels of crude per day in its 42″ (1,067 mm) diameter pipe that was completed in 2005 at the cost of some $4 billion.

Pipelines, once built, are extremely low-cost to operate, due to a combination of being highly automated and a continuous operation. The business case, therefore, needs to be supported by a robust rationale and strong business economics. The current transit tariff for the Baku-Tbilisi-Ceyhan pipeline is just $0.44/bbl.

Petroleum products are loaded in pipelines under pressure, and the product travels at around walking pace. Different products are transmitted through pipelines on a sequential basis in what is known as "*batches*". It is possible to physically separate these batches by use of "*pigs*" or other mechanical devices, but this is rare, so the scheduling of batches and minimizing the number of changes between batches of different products is the art of scheduling pipeline transfers. Very little of sequenced products without a mechanical separator actually do mix; for instance, on a 25,000 barrel batch, occupying 50 miles in a 10″ (250 mm) diameter pipeline, only about 75 barrels would be mixed products. This mix is known as the *interface*, and if the two products sequenced are compatible, then the receiving location would merely downgrade the higher-specification product if (say) premium-grade gasoline was followed in the sequence by regular-grade gasoline.

If, however, diesel fuel preceded or followed a gasoline grade, then the interface mix is incompatible and referred to as a "*transmix*". A series of density measurement devices will detect when the product in the pipeline starts to change from an unadulterated form to the interface and then to a new unadulterated form of the succeeding product batch. The transmix is diverted to isolation tanks to separate this from the main product delivery and usually is returned to the refinery for reprocessing.

Pipeline systems can be extensive, for instance, the USA has over two million miles of pipelines carrying natural gas and hazardous liquids, including petroleum products. Any pipeline that crosses a state line in the USA is defined as a *common carrier,* and its operator cannot refuse to carry and deliver products that meet the conditions and specifications of publicly posted tariffs, which are regulated by the Federal Energy Regulatory Commission (FERC). Pipelines account for around 70% of crude oil and petroleum product movements in the USA. The networks are owned by competing operator systems, so it only costs a few US Cents to move a gallon of gasoline through the 1000 miles (1600 km) from the Gulf Coast to Chicago.

In contrast, the UK's pipeline system is a mixture of Government-developed systems built during the Second World War, and International Oil Company dedicated systems developed in the 1950s and 1960s. The UK network concentrates on the major areas of demand or population, and most systems are operating at capacity — a large proportion of which is the movement of aviation kerosene to London's Heathrow and Gatwick airports, Birmingham and Manchester regional airports.

Capacity productivity can be improved by building pressure booster stations, or reduce the distance between such stations, and the use of friction inhibitors. When such improvements have been explored, the only way to increase product flows between two points is to build another pipeline.

For the movement of petroleum products, crude oil, or natural gas over long distances, in a continual flow, the pipelines provide the only feasible and economical way to transport vast quantities with minimal visual and environmental impact on the land through which the pipeline traverses.

3.7.2 Petroleum product transportation (railcars, coastal tanker, road tankers)

The major petroleum products transported from refineries to end-users and consumers, or indeed intermediate storage terminals, use one of three classic models — coastal tankship, railcars or road tanker. The choice is governed by geography and infrastructure facilities; in a land-locked country, coastal shipping would obviously not be available, and railcar movements require not only a rail network but also loading and discharge facilities.

Railcars are less expensive than road tankers to move petroleum products but can only operate between fixed points of loading and discharge that are physically connected by a rail network. Crude oil and petroleum products are typically conveyed in rail cars, with the advantage that this removes potentially hazardous materials traveling on the road network. Bitumen and ethanol can also be carried in railcars, whereas pipeline transportation of these products can prove problematic. Railcar models are used when longer distances are involved — particularly in the USA, Canada and Russia.

Modern tank cars are usually built to stringent specifications; in the USA, they typically have a maximum capacity of 131,000 L (34,500 US gallons or 28,700 imperial gallons); the equivalent capacities in the UK are 100,000 L (26,300 US gallons or 22,000 imperial gallons). Block trains of 20 railcars (typical UK) would see two million liters moved, or the equivalent of 50 road tanker movements.

The advantages are the reduced environmental impact and less carbon emissions in moving large rakes of tank cars with just one locomotive motive power (Fig. 3.8). Rail car movements are probably ineffective on an economic basis for distances less than 250 km. Another issue may be finding enough "paths" on a rail network where passenger train speeds of 160 kph or greater are the norm. Rail tank cars in Europe are not expected

FIGURE 3.8 A 100-tonne rail tank car (TEA) at Preston, UK. *Used with permission from Paul J. Chancellor, Colour-Rail, (2012).*

to exceed 100 kph. This is less than an issue in Northern America, where most of the rail network is occupied by freight movements only.

Coastal shipping is an essential part of the secondary distribution in island communities, such as the UK. Typically, coastal tankships are found between 3,000 and 10,000 tonnes deadweight (Fig. 3.9). As a modal, speed is slow, and voyages not easily altered once in progress. Less weather dependant than it was historically, as better radar and satellite navigational aids have overcome fog and visibility issues, and enhanced construction overcomes rough seas. Coastal ships are still subject to tidal and berthing restrictions, as much of the infrastructure dates back to the Victorian age in the UK.

FIGURE 3.9 Typical coastal tankship 3500 dwt *Sarnia Liberty.*

Multicompartment tankships can carry a range of different petroleum products on each voyage. Usually, there is a separation between clean and dirty petroleum products, so tankships carrying black oil, or fuel oil are usually dedicated for that purpose so as not to have their tanks cleaned before another voyage when clean petroleum products such as gasoline, diesel or kerosene would be carried.

EU legislation now requires all tankships operating in European waters to have been built less than 25 years ago, and be double-hulled, i.e., the product tank walls themselves are separate to the outer structural hull of the vessel.

Road tanker distribution is invariably the "final mile" of petroleum product distribution to a service station, office, industry, or domestic premises. Road tankers vary in size from small, 4-wheeled rigid tankers, used by distributors to transport products to households, farms, and small industrial premises to large, five- or 6-axle articulated tractor-and-trailer units to deliver fuel to service stations, airports, and industrial complexes.

National weight restrictions, construction-and-use legislation, and build characteristics will determine the maximum payload available. In Europe, where a 44-tonne gross vehicle weight is the usual maximum permitted on public roads, the payloads will be around 40,000 L of gasoline and slightly less for heavier diesel fuel. Most road tankers are multicompartment, which means that three or four different grades of petroleum products can be carried on a single journey.

Where the customer's network is sufficiently large, optimal distribution logistics will see these assets used intensively all year round, on a 7-day week schedule with two driver shifts per truck, per day. Once again, local regulations may limit the amount of driving time on any given day that a tanker driver can work; this may also compromise optimal distribution logistics.

Such intensive usage all-year-round, results in a typical life expectancy of the tractor unit of 6 years, with that of the trailers typically lasting twice that. Capital costs will depend upon specification, but a typical UK 44-tonne tractor-and-trailer unit, as depicted below (Fig. 3.10), would cost around £150,000.

3.7.3 Intermediate storage & terminalling

Often, petroleum products are delivered by road, straight from the refinery to the end user; at other times inland or intermediate terminals are built to break bulk and/or act as a geographical intermediate storage facility (a typical sea-fed terminal is shown at Fig. 3.11). These are often referred to as terminals, and can be supplied with products by sea, rail or pipeline, which are then loaded for onward, and mainly, final distribution to consumers by road.

Optimized terminal operations are usually measured by *"tank turn"* — that is, the number of times in a year that the holding capacity of any tank is completely

FIGURE 3.10 Typical UK 44-tonne articulated road tanker.

replenished. In a well-run and profitable operation, the target for tank turn would be 12 times a year for the major products.

Terminal owners and operators may rent space to product sellers and distributors, usually on a combination of a fixed monthly fee plus a turnover levy, but these financial arrangements can vary enormously between countries, terminalling organizations, and their clients.

FIGURE 3.11 Vopak's Westpoort terminal, Amsterdam, The Netherlands. *Courtesy of Royal Dutch Vopak.*

3.7.4 Economics & optimization

The overriding maxim is to intensely operate the fixed assets, be they pipeline, truck, or terminal. Maximizing the throughput through all these reduces the unit operating costs. It goes without saying that this also has to be carried out safely and with integrity so that the public and the environment are not subjected to undue risk, especially when dealing with highly flammable products such as gasoline and kerosene.

Actual operating costs will vary from country to country, depending on the cost of labor, state regulation, and legislation, but the following table (Table 3.3) does set out the relativity of costs between the transport modals discussed above when related to the movement of petroleum products.

As stated before, not all modals are available in any given location or country; or distance and volumes to be moved will largely determine the most appropriate modal. Therefore, it cannot be assumed that a high-capacity pipeline could actually replace a high-cost road distribution and halve the distribution costs; this solution may be totally impractical, so the table is provided for academic analyses only.

3.8 Refining

Refining is the oil industry's manufacturing heart of transforming the almost-useless properties of crude oil into high-value and useful petroleum products. In the past, the refinery straddled the boundary between upstream (exploration and production) and downstream sales and marketing. At a later date, the refinery was categorized as a midstream operation, which appears to neatly categorize its position in oil's value chain, but with oil company re-organizations that has seen refinery operations managed alongside product marketing and distribution, it is now firmly entrenched as a part of the downstream sector. The Midstream sector now is thought of as defining the transportation and storage of crude oil between oilfields and refineries.

Table 3.3 Model cost comparisons.

Indexed to 100 Pipeline: Low	Low	High	Comment
Pipeline	100	130	Operating costs only
Railcars	135	180	
Coastal shipping	145	350	Low = contract of Affreightment High = spot charter
Road	180	360	Low – 2 loads/shift, 7 days a week High – 1 load/shift, 5 days a week

3.8.1 Basics of refining

Within a refinery, crude oil is processed into more useful product streams by separating the long hydrocarbon chains in crude oil into shorter chains that form the various petroleum products. In the most basic refinery model, this is achieved by heating crude oil in a furnace to over 400°C and then passing the heated crude up a so-called distillation tower. As the heated crude passes up the tower, it cools, and as the hydrocarbon chains break down, the various "raw" product streams separate out. They are captured by a series of trays that allow the upward movement of liquid and vapor, but collect the falling and cooling product, and allow this to be drawn off at various levels in the distillation tower. This process is called ***fractional distillation***.

It, therefore, follows that as the components of crude oil are heated and vaporize, those elements with a lower boiling temperature accumulate at the top of the distillation tower, whereas those with higher boiling points are to be found to separate out toward the bottom of the tower. This quite simple conceptual theory relies on the fact that different hydrocarbons have different boiling points.

Lighter petroleum gases, such as propane and methane, with boiling points less than 40°C collect at the top of the fractional distillation column; next comes gasoline—or more correctly naphtha, which is the feedstock for gasoline—with a boiling point in a range between 25°C and 175°C; this is followed by straight-run kerosene with a boiling point around 200°C, which is the base petroleum product to form jet fuel.

Below kerosene, but included with that product collectively known as ***middle distillate***, is the gasoil and diesel products stream, with a boiling point of around 300°C. The bottom of the distillation tower collects the residue, with a boiling point of 400°C and above. This then has to be further processed, using vacuum distillation to break down the long hydrocarbon chains still further, so that different products, such as lubricant base oils, waxes and bitumen can be produced.

Most petroleum products are subject to secondary and tertiary processes to produce useable fuel for a variety of (mainly) transportation uses. These additional processes include the removal of harmful elements such as lead and sulfur by a hydrotreating unit, reforming processes to improve gasoline performance for higher performance and more sophisticated motor car engines, and by the use of vacuum distillation unit and "***cat cracking***" to produce a greater yield of gasoline and diesel fuel, the more valuable transportation fuels.

A vacuum distillation unit enables more of the longer hydrocarbon chains to be further broken down without the need for further thermal treatment. A catalyst cracking unit ("cat cracker") passes heated gasoil from the distillation column across a bed of fluidized vanadium catalyst into a further fractionating column to produce a greater yield of higher-value petroleum products.

3.8.2   Petroleum products produced

The major petroleum products produced by a typical refinery are listed in order of appearance in the distillation process, from top to bottom, with the lightest products at the top:

- Refinery Gases
- Liquefied Petroleum Gases (Propane, Butane)
- Naphtha (forms gasoline feedstock)
- Kerosene (aviation jet fuel)
- Gasoil (Diesel, GasOil, Marine GasOil)
- Fuel Oils (various blends — Heavy, Medium, Light)
- Residual products (with further treatment, produces lubricants, bitumen, waxes, paints, and solvents)

3.8.3   Refinery yields

Refineries are not like a radio, where you can turn the tuning dial if the music or speech output is not to your taste. Refineries were built and set up to process a certain range or type of feedstock, often referred to as that refinery's crude *slate*. Once this is committed, refineries, especially older ones built in the 1950 and 1960s, cannot easily process another grade of crude oil, just because it is being offered at a discount or is more freely available.

So, for any given refinery, the components that influence and govern its petroleum product output will vary by the age of equipment, the refinery's configuration set-up, and the quality and characteristics of the crude being run. No two refineries are the same; every oil refinery is different in configuration (refer to Box 3.3), and its geographical location may have determined its historical dependence on a particular crude grade. In addition, different and evolving product demands of the country or state in which the refiner sells his products may also influence a required output of a particular petroleum product, which is of less importance in another country, and hence, to that country's refiners.

For illustrative purposes, Fig. 3.12 shows the percentage yields expected from a US refinery before any secondary and tertiary reforming and conversion processes.

Having ineffective personnel, poorly trained, and in an unresponsive organizational structure, will also impact net income by increasing costs, being inefficient, and overseeing the poor operational practice.

All this comes before we consider the marketplace impacts. Increasing environmental legislation has seen petroleum product specifications tighten over time, with politicians and legislation-makers too ready to expect the refiners to invest in additional plant and process to ensure their fuels comply, without assisting in the capital expenditure investment to do this. The refiner will not sell an additional barrel of product to help defray this expenditure, but shareholders will be expecting dividends and earnings to be nevertheless maintained.

BOX 3.3 Insight: refinery categories and complexity

TOPPING REFINERIES are the most basic of refineries and uses just a simple distillation process, either just an atmospheric distillation or this plus a vacuum distillation tower. The output of such refineries is determined by the crude oil run, and the products produce by these are often half-finished that have to be further processed in other refineries.

HYDRO-KIMMING REFINERIES have reforming units which produce hydrogen that can be used to increase the octane rating of the gasoline stream as well as lowering the sulphur content of distillate fuels.

CRACKING REFINERIES have a catalytic cracking unit which also increases the gasoline yield and minimises the production of lower-value products from the bottom of the distillation column. A variant of this category can be coking refineries which produce petroleum coke from the residual product at the bottom of the distillation process.

COMPLEX REFINERIES are coking refineries with even more processing units, such as a steam cracker that can produce feedstock for the petrochemical industry. In essence, a full conversion refinery is really a combined oil refinery and petrochemical plant ion one processing site. Complexity increases with every additional processing unit that is added to the basic atmospheric distillation tower. Refineries have become increasingly complex over time, partly driven by increasing gasoline demand fuelled by the growth in automobile ownership, bigger refinery units being built in India and the Far East, and a business case driver of built in future-proofing to be able to run a wider slate of crude oil than those refineries built for one specific crude oil, built in the 1950s and 1960s.

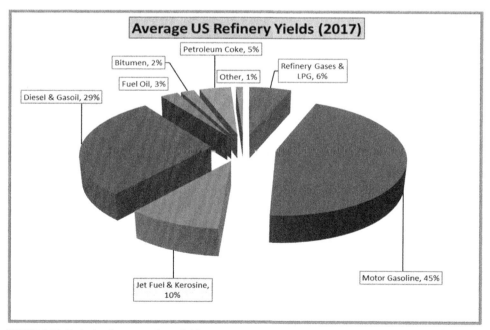

FIGURE 3.12 Typical refinery product Yields —USA. *Data: US Energy Information Agency; Graphic: Author.*

3.8.4 Challenges

If we considered a refinery as a stand-alone business, then aside from the inability of the refiner to control the costs of both his feedstocks (crude oil) and his outputs (petroleum products) that are subject to commodity market fluctuations far beyond his sphere of influence, there are myriad challenges from economic to environmental and safety aspects.

Obviously, from a business perspective, maximizing income and minimizing costs are paramount. But while operating the refinery to produce both a net income and [say] a 15% Return on Capital Investment, the refiner must ensure the integrity of day-to-day operations where neither the environment nor the workers are harmed, endangered and kept free from risk. In addition, their operational integrity must meet the local legislation on health, safety, security, and environmental matters.

It is somewhat easier to fail than to achieve a flawless operation. Acquiring the crude oil feedstock at a less-than-optimal cost, or having to sell refinery products ex-jetty as a distress sale to create additional storage capacity to accept newly-delivered refined products will severely impact net income. Refineries operate most economically when in continuous operation at near-to-capacity volumes. Shut-downs for maintenance need to be scheduled carefully, and turn-round times minimised whilst ensuring quality and safety standards are assured.

Having ineffective personnel, poorly trained and in an unresponsive organisational structure will also impact on net income by increasing costs, being inefficient and overseeing poor operational practice.

All this comes before we consider the marketplace impacts. Increasing environmental legislation has seen petroleum product specifications tighten over time, with politicians and legislation-makers too ready to expect the refiners to invest in additional plant and process to ensure their fuels comply, without assisting in the capital expenditure investment to do this. The refiner will not sell an additional barrel of product to defray this expenditure, but shareholders will be expecting dividends and earnings to be nevertheless maintained. As an example of this, the European Union now demands that the sulphur content of road transportation fuels does not exceed 10 mg/kg — often expressed as 10 ppm[5]; whereas in 1999, the permitted limit was 500 ppm. These tightening regulations were the result of the EU's Auto Oil program, designed to improve air quality. Just this one aspect causes refiners to invest in additional process units for hydro-treating to remove and recover the sulphur from road fuels and additional fractionation to remove excess aromatics and benzene in gasoline.

BP's Statistical Review of World Energy (2018) reported that refinery utilisation globally was 83.7%, the highest figure for 9 years. However, this global average masks some widely divergent regional analysis; South & Central America refineries showed a 66% utilisation rate and Africa around 50% for the past 3 years. In contrast, the other regions, including Europe and North America has utilisation figures in the range

[5]ppm: parts per million.

between 80% and 90%. It would be unusual for refineries to run consistently above 95% utilisation; as such over-use will probably result in bottlenecking in some other part of the refinery site. An old axiom in the oil and gas industry is to maximise throughput through any infrastructure asset, and to keep operating on a 24/7 basis, thereby reducing the per unit operating cost to its lowest possible figure. Refineries are no exception to this rule.

3.8.5 Process safety

Process safety is the key to managing operational risk; where this fails, major disasters such as Deepwater Horizon, Buncefield, Texas City refinery, Alpha Piper, and Jaipur occur. These incidents build upon each other in the minds of the public and politicians, with individuals called to account and no longer protected by the collective corporate responsibility. This leads to increased scrutiny through regulation and compliance orders, which, in turn, can undermine investor confidence in refineries and refiners, irrespective of whether the operation is making good financial returns.

Predicting whether tomorrow will be an incident-free day is the challenge, and which begs two further questions of how this will be achieved and how we can measure this achievement.

Achieving that flawless refinery operation, to achieve operational excellence is probably not going to happen without the use of a conceptual model or structural framework. From the identification of risk and assessing that risk for probability and impact, through risk mitigation measures, to a constant review and improvement process would be the basis of such a conceptual framework.

Using a framework or structure will permit the measurement of how well the refinery is doing on an absolute scale and thereby identifying areas for concern within a process safety profile. By benchmarking within the refinery's peer group, a relative measure can also identify just how well the safety and operational integrity of one refinery is doing against the rest of the industry.

Process safety has emerged as a higher-profile concern for refiners during the past decade, and the questions have developed from just recording incidents into ensuring the integrity of the operation, and also proving that the refiner knows that this is being done.

3.8.6 Refining economics

Several factors, as have been previously discussed, can affect refinery economics and profitability. In this section, these will be discussed in greater detail.

Refining is a capital-intensive business, and in oil companies, the refining section has to compete with other operating decisions to gain the capital expenditure authorization to make any required upgrading investments, even if these are merely to keep pace with changing legislation and tightening product specifications.

As stated before, the cost of crude oil and any other feedstocks such as naphtha will have prices set by the market, as is normal in globally traded commodities. This is perhaps the major variable cost in the refining process, and the sometime-volatility in prices will bring about rapid up-and down-swings in refinery profits.

Labor costs in a refining operation are but a small portion of the overall costs but can attract a disproportionate share of management analysis and review because they are eminently controllable by the refiner.

On the assumption that the refinery is producing on-specification fuels, then demand will be set by the market in the long-term; prices for products are just the same as crude oil, another commodity, traded globally, and outwith the control of the refiner, who has precious little opportunity to introduce any product differntiation or brand loyalty, so as to be able to demand a premium on top of the commodity product price.

It does not necessarily follow that low crude oil prices have an adverse effect on refinery profitability; the reverse is also true, as evidenced by the graphic (Fig. 3.13) reproduced below:

As can be seen above, aside from the correlation between high crude prices and high refining margins between 2003 and 2006, the analysis of the remaining years disavow the production of any discernible trends or alignment.

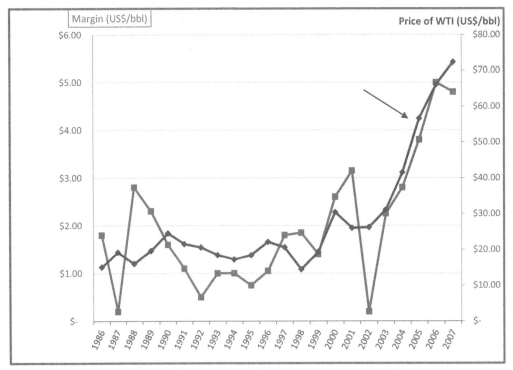

FIGURE 3.13 US refining margins ($/bbl) versus WTI Crude price ($/bbl). *Source: US Energy Information Administration.*

For fully vertically integrated oil companies, choosing whether to invest in their refining sector, there are always competing investment opportunities. Some idea of the scale of these pressures can be discerned from the following graphic (Fig. 3.14) that vividly demonstrates lower crude oil prices adversely affecting their upstream earnings. It is salutary to note, however, that in the previous four years prior to 2015 that downstream earnings, which include refining, never exceeded 30% of all globally integrated oil company earnings.

The US refining sector reported negative net earnings of some $7 billion in 2009, and the Return on Investment is perhaps at its lowest level ever reported at minus 7%. The following graph (Fig. 3.15) illustrates the contrast of these returns against the immediately preceding 5-year period, which saw unprecedented profits earned by US refiners. Outside the USA, the results are better, but only in comparison. Refining economics remain under severe, and cyclical, pressure with 5% returns inadequate against the 15% benchmark we have suggested is needed for flawless refinery operation. The period during the 1990s and early 2000s saw successful cost-cutting programs, leading to increased profitability, but since then, petroleum products have fallen faster than crude oil input costs, with the inevitable squeeze on refinery earnings and returns.

As can be seen from the following graphic (Fig. 3.16), analysis by the US Energy Administration has shown a strong convergence of refining earnings. Historically North American refineries were consistently more profitable than their counterparts in Europe; now, this differential has been largely eroded, the reason being the changes in the differentials between regional crude oil costs. The narrowing of the crude price differential is driving the convergence in earnings, and crude oil prices (feedstock input costs) are a primary driver of refinery profitability. Since the latter part of 2015, the differential between North Sea Brent crude and (say) West Texas Intermediate has not breached $4.50 per barrel, thereby eroding the previous cost advantages held by US refiners.

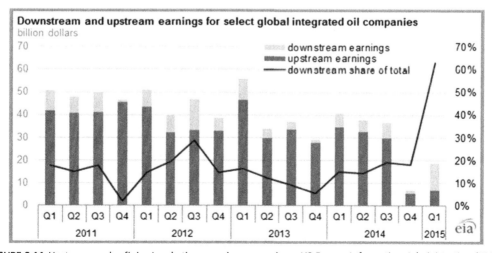

FIGURE 3.14 Upstream and refining/marketing margins comparison. *US Energy Information Administration (2011).*

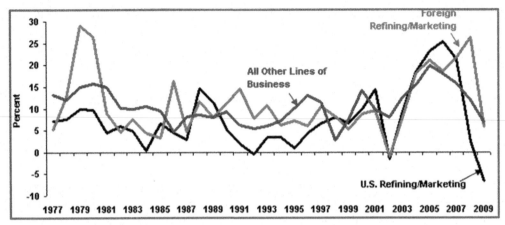

FIGURE 3.15 Return on investment for US and foreign refining/marketing 1977–2009. *US Energy Information Administration (2011).*

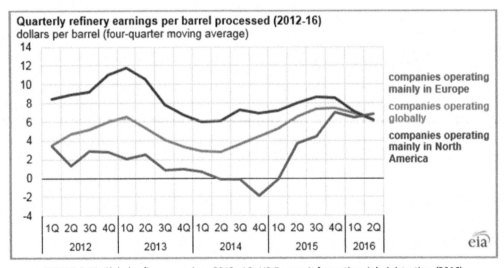

FIGURE 3.16 Global refinery earnings 2012–16. *US Energy Information Administration (2016).*

For an added comparison, UKPIA[6] stated in their *2018 Statistical Review* that the return on investment for its members in the years 2012–2016 was around 9%. In the same time period, their analysis shows this in comparison to returns of 13% in three comparable industries, being 16% in the service sector, upstream 13%, and that for manufacturing 9.5%.

The challenges for refiners are not going away. Maintain operational integrity, a fully safe operation, mitigation of risks, and having to comply with increasing environmental legislation and tightening product specifications; all are having to be achieved with single-digit returns. Their financial outlook remains bleak, but as the oil industry cannot

[6]UKPIA: United Kingdom Petroleum Industry Association.

exist with refining creating valuable petroleum products on which 21st Century life depends all over the world, the industry has no choice but to persist with low returns and high investment models and frameworks for the refining sector.

3.9 End users

3.9.1 Aviation

Kerosene has been the chosen fuel for the aviation industry due to its wide availability and excellent energy content per weight, an important consideration when the aircraft has to lift and carry its fuel source for very long journeys.

Quality of jet fuel is a most important requirement, and the care of kerosene batches from production at a refinery all the way through the supply chain and infrastructure and into the fuel tanks on an aircraft is carefully tested and maintained at all stages. The industry embraces the needs of this utter transparency, and traceability has meant that aircraft accidents due to off-specification aviation fuel are extremely rare.

While a high degree of redundancy is built into aircraft engines, fuel quality is of obvious importance. With 40 million flights scheduled each year, this translates down to operations involving the refuelling of some 250,000 aircraft each and every day.

Airlines rate the fuel quality and characteristics above price competition. These additional product qualities include thermal stability, lubricity, fuel system icing inhibitors, antioxidants, and detergents. Only additives specifically approved can be added to jet fuel; individual oil companies, therefore, cannot claim differentiation of jet fuel from that of another firm by the inclusion of different additives.

There are a small number of specifications that are used globally - see Box 3.4 below:

Standards for the detail of aviation fuel quality control and fuel handling procedures are promulgated by the Joint Inspection Group (JIG), which includes Guarantor Members: Eni, BP, Chevron, ExxonMobil, Kuwait Petroleum, Shell, Statoil, and Total. In addition, the Energy Institute (EI) of the UK published in 2013, a joint EI/JIG Standard 1530 which is:

> " ... intended to provide a standard to assist in the maintenance of aviation fuel quality, from its point of manufacture to delivery to airports. It provides mandatory provision and good practice recommendation for the design/ functional requirements of facilities, and operational procedures."
>
> (Energy Institute & Joint Inspection Group (2013))

BOX 3.4 Insight: aviation jet fuel — global standards

❖ **Jet A** — used at airports throughout the USA and in certain parts of Canada; usually contains no additives
❖ **Jet A1** — used at airports outside of North America, developed by oil companies specifically to assure airlines operating globally, and refueling in a variety of countries. A similar fuel to Jet A1 is used in China.
❖ **Jet TS-1** — used within the countries making up the former Soviet Union

3.9.2 Road transportation (hauliers, railways, shipping)

Diesel fuel, in one form or another, powers the engine for most forms of transport on land and sea. This middle distillate fuel can be found described in a variety of ways, and that differentiation either refers to a specific use — marine gasoil — or a different taxation treatment by the Government. For instance, diesel gasoil for use in a vehicle that uses public roads is likely to be taxed at a higher rate of *specific unit tax* than (say) the same equivalent fuel for agricultural use in tractors.

In the UK, diesel fuel for use on public roads is colloquially known as DERV (Diesel Engine in Road Vehicles) to distinguish it from gasoil for agricultural and off-road use. There could be some product differential characteristics, but the reason for this separate classification is that the Excise Duty on UK Derv is some five times higher than that levied on gasoil. In order to prevent fraudulent use and to aid detection, gasoil will be dyed red, and both fuels will contain chemical markers that can be detected and analyzed in a laboratory. In Europe, Solvent Yellow 124 has been the dye used since 2002 to mark and dye agricultural and heating diesel (gasoil) from that diesel (DERV) used in road vehicles.

Railways all over the world turned to the use of gasoil diesel to power locomotives that replaced steam-powered locomotives from the 1950s onwards. Diesel propulsion is being replaced in large scale electrification across Europe and the Far East, but the capital cost of the infrastructure and reliable electricity supplies are not always available, and thus, diesel-powered locomotives and passenger trains will continue to be a major consumer of distillate fuel.

Historically, deep-sea ships have burnt fuel oil for propulsion, but increasingly tighter fuel specifications as mandated by MARPOL [International Convention for the Prevention of Pollution from Ships] to which over 99% of the world's ship owning countries belong, are being introduced to reduce harmful emissions from ships. This has already seen the sulfur content of marine fuels being burnt reduced to 0.10% in the Sulfur Emission Control Area (SECAs) of the Baltic Sea, North Sea, and most of the USA and Canadian coast. All other shipping will be forced to ensure that emissions are reduced by using fuel with lower than 0.50% sulfur by weight; this will require moving from fuel oil for propulsion to marine gasoil similar in characteristics to ordinary gasoil for agricultural and road use. If ship operators wish to continue to burn the much cheaper residual fuel oil, then their ships will have to scrub the exhaust before it is emitted into the atmosphere.

3.9.3 Service station networks

Although service stations are not really end-users, they are the last link in oil's value chain as far as motorists are concerned. In the developed world, service stations immediately conjure up visions of large, oil company-branded outlets, but there are many different channels to market that service stations can take. In many African countries, roadside stalls can sell gasoline in jars and bottles; in France grocery hyper-markets dominate fuel retailing; in the USA there are many different channels; in government-controlled markets, NOCs control the retail network.

Facilities and throughputs per service station vary widely across the globe but will reflect the evolutionary history of the particular retail network. Refiners have, in the past, established their own network of company-owned service stations to provide a "locked-in" outlet for their refinery output. The ultimate stage of this integration was to see refiners manage and operate their company-owned network, such that fully vertically integrated international oil companies controlled the whole of the oil value chain from wellhead to petrol pump.

This level of vertical integration has largely disappeared, a response to the challenge of greater competition and falling margins. Specialist retail operating companies can now provide the expertise and assure IOCs of quality control, such that the IOCs can outsource much of their operations, yet retain their public branding and thus public consciousness, and concentrate on their core operations – usually exploration and production.

IOC and NOC refiners can also sell retail petroleum products into the spot market or to large retail groups, which may include supermarket and hypermarket chains. This wholesale supply sees products sold to competing businesses that then compete with IOC and NOC products sold through their own service station networks.

In developed markets, service station networks will be owned and operated by a complete mixture of ownership and operating models, such that any one station could be owned by an oil company, a supermarket, a distributor (known as a jobber in the US), or an independent dealer.

The range of facilities at service stations will also greatly vary. Due to inadequate returns form the process of just selling fuels, costs have been squeezed out of the operation by conversion to self-service operation and pay-at-pump dispensers, and the development of convenience stores to provide additional income streams at the service station with margins that are less competitive, and thus, provide a more sustainable profit and return for the service station operator.

The display of an oil company branding does not guarantee that the fuel dispensed at that site is actually refined by that oil company and that the fuel is supplied from that oil company's refinery. In order to effect optimal distribution and logistics, oil companies will exchange or swap products, whereby the most cost-effective supply route, distance and cost are used. This has evolved over time and replaced much unnecessary duplication of infrastructural assets in the supply chain. Greater critical mass can be obtained by sharing terminals and pipelines to produce the supply chains extant today.

To assure motorists that quality of fuel is never compromised by the myriad of supply arrangements and companies involved upstream of the petrol pumps themselves, national standards and specifications ensure all fuels retailed meet a minimum quality requirement. These standards are usually set by governments, but in conjunction with the motor manufacturers and the oil companies have evolved as the understanding and science of the harmful component within gasoline and diesel, and their detrimental effects on the environment and public health, have led to tighter environmental specifications for both engines and fuel. This is evidenced by the removal of lead from gasoline, where it was once used to improve octane ratings; and sulfur from all road fuels.

These national specifications create a benchmark for all refiners and operators in that country that must be met, and perversely encourage further competition, as lower prices — such as supermarket fuel prices — do not mean lower quality.

3.9.4 Heating oil (Domestic and industrial)

The particular petroleum product used for heating domestic premises varies across the globe. Natural gas — historically from the North Sea, and now increasingly Russia — is extensively used throughout Europe. In the UK, off-grid heating is provided by kerosene, with individual properties having their own storage tank. In Germany, for instance, much heating is provided from centralized systems with large tertiary storage for the heating oil, which is gasoil, a middle distillate fuel. Heating oil has historically been the fuel of choice in eastern Canada and the north-eastern USA, where there was limited access to natural gas.

Industrial heating was often provided by burning gasoil, or by generating steam from burning fuel oil and then pumping the resultant hot water through radiator systems. This has been largely supplanted by natural gas systems and even electricity. The move away from heavy industrial processes into service-based economies, housed in offices rather than factories, has accelerated the decline in industrial heating oil demand. This lessening of demand can be evidenced by the reference to the UK Government's Digest of Energy Statistics for 2017, in which industrial gasoil demand represents just 2.7% of overall final consumption (compared to a 75% share for the three transportation fuels of gasoline, jet kerosene, and diesel fuel) (HM Government, 2018).

3.9.5 Other users/uses (petroleum gases; lubricants; waxes; bitumen)

Liquefied petroleum gases (LPG) such as ethane, propane, and butane, along with refinery gases, will only account for around 5% of a refinery's typical yield from crude oil. Ethane is the lightest of the natural gas liquids and finds application majorly as a feedstock for petrochemicals, for ethylene production, and to a minor extent as a refrigerant gas. In addition, ethane can be combined with benzene to manufacture polystyrene food containers, and with chlorination treatment can produce PVC (plastic polyvinyl chloride). As a refrigerant gas, ethane is easily compressed and then absorbs heat as it expands in the walls of a refrigerator.

Around half of the propane produced finds its way into the petrochemical industry manufacturing plastics, and thus, has to compete with other LPGs and naphtha on price. A significant proportion of propane is used for domestic and commercial heating purposes; a small proportion of propane is used in automotive applications as an alternative fuel to gasoline (often known as Autogas). Propane prices reflect movements in both natural gas and oil prices. Some propane demand is obviously seasonal, and is, therefore, stored in underground salt caverns during low demand in summer, and released in the winter to meet the increased heating demand.

Normal butane is found in both domestic LPG (around 10% component share) and as a cigarette lighter fuel. As it possesses a higher octane rating butane can be directly mixed into gasoline production as an important blendstock. As it is a volatile gas, butane is more useful as a gasoline blendstock in cold winter months than summer months.

Refinery Base Oil outputs are used to manufacture lubricants and greases, in turn used in end applications to lessen friction in plant and machinery, where the end-users' requirements are for lubricants that do not prevent starting engines on cold days because the lubricating oil used has thickened in viscosity, while also not thinning out at high operating or ambient temperatures, and thus losing the protection against the forces of friction.

Waxes are hydrocarbons that exist in a solid-state at normal ambient temperatures, but which can also be heated many times to reform into articles such as candles, waxed paper, and a variety of food-related waxes. End-user products that are made from waxes will include chewing gum, petroleum jelly (Vaseline), and paraffin wax (candles).

One of the heaviest refinery products is bitumen, which must be heated above 150°C to maintain its liquidity, and is used in asphalt manufacture for road surfacing. This production is seasonal, with no real demand in the winter months, and representing around 85% of all bitumen produced at the refinery. Tar is sometimes confused with both bitumen and asphalt but is actually a by-product of coal and not oil.

References

BP, 2018. BP Statistical Review of World Energy 2018. BP, London.

Commodity Research Bureau, 2018. The CRB Commodity Yearbook 2018. John Wiley & Sons Ltd, New York.

Energy Information Administration, 2011. eia.gov [Online] [Accessed 2 November 2018].

Energy Information Administration, 2016. eia.gov [Online] [Accessed 2 November 2018].

Energy Information Administration, 2018. eia.gov [Online] [Accessed 2 November 2018].

Energy Institute website: https://www.energyinst.org/home.

Energy Institute & Joint Inspection Group, 2013. Quality assurance requirements for the manufacture, Storage and Distribution of Aviation Fuels to Airports, 1st ed. Energy Institute, London.

GeoExpert, 2019. (online) available at: https://geoexpert.ch/methods/reflection-seismic-surveying/ [last accessed 12 August 2020].

HM Government, 2018. Digest of UK Energy Statistics (DUKES) [Online] Available at: https://www.gov.uk/government/collections/digest-of-uk-energy-statistics-dukes [Accessed 21 October 2018].

Joint Inspection Group (JIG), website: www.jigonline.com.

Meriam-Webster Dictionary, 2018. website: http://www.merriam-webster.com.

OPEC, 2018. OPEC website [Online] Available at: https://www.opec.org/opec_web/en/about_us/23.htm [Accessed 28 October 2018].

Oxford Living Dictionaries, 2018. [Online] Available at. http://languages.oup.com.

Further reading

Association for Project Management, 2004. [Online] Available at: www.apm.org.uk https://www.apm. org.uk/body-of-knowledge/context/governance/project-management/ [Accessed 11 November 2018].

Atrill, P., McLaney, E., 2016. Financial Accounting for Decision Makers, 8th ed. Pearson Education Limited, Harlow, UK.

Badiru, A.B., Osisanya, S.O., 2013. Project Management for the Oil and Gas Industry: A World System Approach. CRC Press, Boca Raton, Florida, USA.

Canadian Energy Research Institute, 2017. Canadian Oil Sands Supply Costs abnd Development Projects (2016-20136. CERI, Calgary, Alberta.

EI/JIG, 2019. EI/JIG Standard 1530 Quality assurance requirements for the manufacture, storage and distribution of aviation fuel to airports, 2nd ed Energy Institute, London.

Herkenhoff, L., 2014. A Profile of the Oil and Gas Industry. Business Expert Press, LLC, New York.

James, T., 2008. Energy Markets: Price Risk Management and Trading. John Wiley & Sones (Asia) Pte. Ltd, Singapore.

Merrow, E.W., 2011. Oil and Gas Industry Megaprojects: Our Recent Track Record. Society of Petroleum Engineers, Houston, Texas, USA.

Natural Resources Canada, 2014. Energy Markets Fact Book 2014-2015, s.l. National resources Canada.

Natural Resources Canada, 2017. Oil Resources [Online] Available at. www.nrcan.gc.ca/energy/oil-sands/18085 [Accessed 12 October 2018].

Project Management Institute, 2018. Project Management Body of Knowledge [Online] Available at: https://www.pmi.org/pmbok-guide-standards [Accessed 11 November 2018].

United Nations, 2017. World Population Prospects: The 2017 Revision [Online] Available at. https:// population.un.org/wpp/Download/Standard/Population/ [Accessed 21 October 2018].

4

Natural gas

Gas has often been found together with oil deposits; the quality of the gas found, its distance from consuming markets would have historically determined whether the gas deposit was commercially exploited. Gas is considered to be the most environmentally friendly fossil fuel, which is not nowadays seen as a ringing endorsement.

Demand for natural gas comes from three sectors, residential (or domestic) and industrial and commercial demand for gas for heating and cooking; for power generation in turbines generating electricity from burning gas; and in natural gas-powered vehicles that use Compressed Natural Gas.

4.1 How gas is formed

Gas is formed by the same geological process as for oil, namely, the crushing and decomposition of organic matter deep underground and near the earth's hot core, as higher temperatures and greater pressures are usually required to form the lighter hydrocarbons such as natural gas. Typically, gas associated with oil will be found at the top of conventional oil reservoirs found under anticlines (or domes) trapped from migrating further upwards by layers of impermeable rock.

When wells searching for oil were drilled, gas was often encountered—this was known as associated gas—and as this gas was not commercially exploitable using the then-known technology, it was regarded as an unwanted by-product of the oil exploration and was thus burnt, or "flared," off at the well-head. This was a waste of energy, and if associated gas is discovered today, it is reinjected into the oil reservoir to aid further recovery of the oil from the reservoir. Associated gas could further be delineated as either free or dissolved gas, free gas separated out from the oil while below ground; dissolved gas often separates out during the process of pumping the oil to the surface.

Natural gas is predominately methane (chemical formula CH_4), with up to 20% of propane (C_3H_8) and ethane (C_2H_6). Gas had a 23% share of all primary energy consumption in 2017 (BP, 2018). It is colourless and odorless that must be injected with a stenching agent, mercaptan, so that any leak can be detected by smell.

Condensate gas wells produce a semi-liquid hydrocarbon called condensate, which is almost pure gasoline, which is of considerable value when present. Natural gas that is almost pure methane is known as Dry natural gas; when condensate and other compounds are present, over and above just methane and ethane, then the gas is known as Wet natural gas. Sour natural gas contains larger amounts of hydrogen sulfide, which is highly corrosive and would have to be removed before the commercial application of the gas.

Rural Electrification. https://doi.org/10.1016/B978-0-12-822403-8.00004-7

4.2 Geography

By comparison with the oil industry, the gas sector is relatively immature in its evolution, in spite of the fact that associated gas has been present in oil finds for many decades. The advent of LNG and its associated overturn of the transportation challenges will begin to improve the connectivity between the geographical markets and the supply regions.

Like oil, the geography of natural gas shows a wide disconnect between gas reserves and production and those countries in which it is consumed. The Western Hemisphere does not contain natural gas resources, which are concentrated in Asia and the Middle East, as the following graphic (Fig. 4.1) demonstrates.

The largest natural gas field is offshore of Qatar and Iran and is thought to contain more than 1200 tcm,[1] dwarfing all other gas fields currently discovered. Production, even at the optimum rate, would not deplete the gas field in 200 years.

Gas production and development in the past ensured that consuming markets emerged only if in close proximity to gas reserves, and thus, the current geography of consumption differs widely from the emergence of these new deposits, as this graphic of consumption demonstrates, see Fig. 4.2.

4.2.1 Lack of historical interconnection between gas fields and consumers

The natural gas market historically is, by comparison with the global oil market, at best a regional-based set of individual markets. The US market focusses on Henry Hub, a major

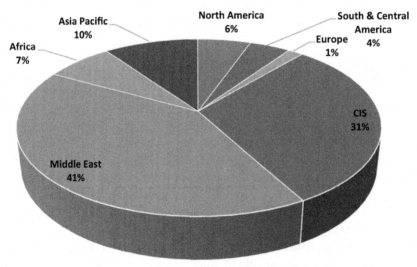

FIGURE 4.1 Proven natural gas reserves. *Data (BP, 2018); Graphics: Author.*

---

[1]tcm = trillion cubic meters of gas.

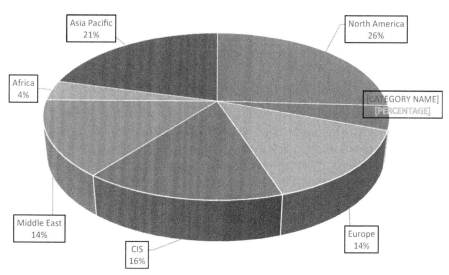

FIGURE 4.2 Natural gas consumption. *Data (BP, 2018); Graphic: Author.*

pipeline interchange situated in Earath, Louisiana, which is the benchmark delivery point for the physical delivery of gas on expiring gas futures contracts. The North American market is a very well developed and liquid spot market for natural gas. Henry Hub's position as a price discovery indicator of spot market trading and pricing for natural gas is justified by wide-scale third-party access to the hub. Gas prices and crude oil supply prices are inherently separate markets in North America.

The advent of both Shale Gas production in the USA and the ability to convert Natural Gas into Liquefied Natural Gas (LNG) and thus a transportable liquid fuel, will increasingly connect gas supplies by LNG ships from the eastern seaboard of the USA to the UK. In turn, the UK gas market was born out of replacing town gas made from coal by naturally occurring North Sea gas discovered close enough to make commercially viable the shipping of gas ashore by pipeline to the UK.

Additionally, the UK gas market has a physical connection to the European market through several pipeline connections (Table 4.1 below), and especially imports a large amount of natural gas from Norway through the Langeled pipeline from Nyhamna in Norway to Easington in Yorkshire, and by the Vesterled from a number of Norwegian gas fields to St Fergus in Scotland. The two other pipeline connections are the UK–Belgium

Table 4.1   UK natural gas pipeline connections with Europe.

Pipeline	From	To	Capacity (bcm per year)
IUK	Zeebrugge, Belgium	Bacton, Norfolk, UK	25.5
BBL	Balgzand, Netherlands	Bacton, Norfolk	14.2
Vesterled	Norwegian gas fields	St Fergus, Scotland	14.2
Langeled	Nyhamna, Norfolk	Easington, Yorkshire	26.3

Interconnector from Zeebrugge in Belgium to Bacton in Norfolk (this is the only bidi-rectional pipeline and can thus *export* UK natural gas to Europe), and the UK–Netherland pipeline from Balgzand to Bacton.

4.2.2   Challenges of storage and transportation versus oil

Oil has the obvious advantage over gas in that, as a liquid, it can be easily contained, stored, and transported. Transportation of oil by any of the modals was never an issue from the early days of wooden barrels and pipelines in the 19th Century. An inability, for much of the 20th Century' to be able to convert natural gas into a transportable liquid meant that the commercial exploitation of natural gas required the gas field to be in close proximity to consumers with pipeline connections.

The advantages of LNG as a fuel that can bridge the enormous distances between the newly developing gas fields and the historical consuming markets will be discussed later in this Chapter under 4.3 *The Market for Natural Gas* and 4.4 *LNG Transportation for Natural Gas.*

4.3 The market for natural gas

4.3.1   Market regionalization

Historically, natural gas was flared off in the oil field when it was generated almost as a waste by-product of oil production. As this was realized as a profligate waste of energy, natural gas was transported relatively short distances by pipeline to supplant local generation of town gas from burning coal in furnaces, which fed domestic distribution grid networks. While this was a less wasteful energy usage, it could only prove efficient where the consumers could be reached by pipeline distribution modals direct from the production gas field.

So during oil's evolution, when technical solutions were solved to transport large quantities of crude halfway across the globe, gas became more deeply entrenched into a regional market structure, focused on North America, Europe, and the Pacific. At the same time, oil was being traded on a truly global basis because solutions to store and transport oil and petroleum products had been developed, improved, and enhanced.

The North American gas market is built upon a network of pipelines crossing this continent, and the market was developed on the back of historical growth in the US oil market that saw large quantities of associated gas produced that drove the rapid development of these pipeline systems. Europe experienced a period of rapid growth in the 1970s and 1980s when the North Sea production basin was at its peak, and its subsequent decline was mitigated by the growth in Russian gas exports. Most of this gas reaches import terminals in European coastal countries by sub-sea pipelines. Growth in consumption since the Millennium has seen declining North Sea gas production being supplanted by LNG imports from places such as Qatar and Australia.

The Pacific market is dominated by Japan, which relies almost 100% on imported energy. The proximity of natural gas in Indonesia and Australia has led to an obvious development of an LNG-import market.

The technology to cool natural gas to a sufficiently low enough temperature so as to allow the transportation of liquefied natural gas (LNG) has only been successfully commercialized since the late 1950s and the beginning of truly global trading markets in LNG has now started to emerge. The gas market has some considerable catching up to match oil's interchangeability of crude from oilfield to refinery, the ready availability of spot cargo trade and oil's liquidity, and transparency of price.

4.3.2 Integrated supply chain for natural gas using LNG trains

LNG has only recently provided the natural gas sector with the technology that has overcome the obstacle of the transportability issues involved with moving natural gas between continents over vast sea distances. Achieving this global competitiveness comes at a huge price.

The capital expenditure levels required, which make up around 80% of the total liquefaction costs, and the cost of LNG extraction often result in fully integrated arrangements that commit the gas itself, the LNG liquefaction plant, the specialist tankships, and the import terminal and regasification facilities into one contract arrangement. These are often for a 20- to 30-year period, this length required by the level of amortization required for the vast capital sums involved. In addition, end-use customers, such as gas distributors and power generators, are often locked-in by lengthy *take-or-pay* contracts. It is believed that to support such an integrated contract approach to LNG demands that the gas field must support the production of at least one million tonnes for 20 years.

Gas processing and storage units at the liquefaction end of the chain are known as "*Trains*" and can cost in excess of $1500 per mt in capital expenditure, and the operation of the train will account for half of the operating costs of the whole supply chain.

Regasification plant building capital costs are of the order of $1 billion per billion cubic feet of gas per day capacity, even though the process is much simpler than liquefaction. Overall, LNG remains the most expensive fossil fuel to transport to market, twice as expensive as gasoline and three-and-a-half times that for crude oil.

There are two major markets for LNG, Asia–Pacific, and the Atlantic Basin. The larger of these is Asia–Pacific, where Japan is the single biggest consumer with supply originating from Indonesia or Malaysia, which are still some significant distance from the large consuming countries. This distance, together with the fact that LNG prices are based on a Japanese crude oil cocktail pricing formula, ensures that this region endures the highest LNG prices in the world. Most of the trade is carried out under long-term contracts.

In Europe, historically, most of the LNG supply has originated in North Africa, but this is changing with new import channels from Qatar, the Far East, and even Australia. Pricing of LNG is complex, reflecting traditional gas pricing by region rather than a European or global benchmark. The UK tends to use pricing at the National Balancing Point (NBP), which is a virtual trading location, and also is the pricing and delivery point for ICE natural gas futures contracts. At the NBP, gas is traded in units of pence per therm.

The Henry Hub is a physical receipt and delivery location in Erath, Louisiana, and is the pricing point for New York Mercantile Exchange's (NYMEX) natural gas futures contracts as well as OTC swaps traded on the Intercontinental Exchange (ICE) in London. Its significance as a physical trading hub is the interconnection with many interstate and intrastate gas pipelines. Gas prices at Henry Hub use a denomination of $/MMBtu ($ per millions of British Thermal Units) and set the prices for the whole of North America.

4.3.3   Developing the LNG market in contrast with that for oil

LNG is slowly revolutionizing the natural gas industry, and the evolution in both the required technology and the LNG tankship design are propelling natural gas from a very tenuously-connected three regional markets into a more unified global market, using LNG transportation as the primary driver to making natural gas a more transportable global commodity.

LNG mega-projects require billions in up-front infrastructure development, and in order to thus develop this infrastructure to bring more liquidity and product availability to the market, multidecade project and supply chain commitments are required from partners and joint investors. Oil is at a stage of its 150+ year evolution, where investment cycles are much shorter, and therefore, will continue to offer more flexible opportunities than natural gas and LNG.

If this evolution for natural gas follows the precedents set by oil, then the establishment of tradable LNG benchmarks must follow, supported by market liquidity in spot trades and price discovery. The threat of a "gas troika"—an equivalent to OPEC's cartel—would be a disaster for gas consumers, as restricted or controlled supplies would not allow free competition and thus floating prices. The gas troika has been discussed for over a decade between Russia, Iran, and Qatar, which had a combined hold over 50% of the world's gas reserves.

As natural gas is not yet sold on a global market, and the transportation issues are not as easily solved as they are for oil, it will be harder to establish a cartel to control gas prices and supply, aside from the differences in their political issues and aims that these three countries have. The debate about a troika is set to continue.

In contrast to the oil market, much of the natural gas market deals with "*stranded*" assets, meaning that the gas fields have to be connected by pipeline infrastructure to consumers, in turn, necessitating long 25-year contracts for amortization of expensive capital works. This fact alone defeats the evolution of a widely traded, fungible spot market to sell off a surplus or supplement a shortfall in supply. LNG transportation could overcome the problems with these stranded assets, but until such time as the infrastructural assets have been fully amortized and there is sufficient spare capacity in transportation, storage, and distribution assets, spot market trade in LNG will effectively be squeezed out by companies that have invested in an integrated LNG supply chain as the only way to guarantee a return by committing all players at every link in the chain. That day will surely come, but not any time soon.

4.4 LNG Transportation for natural gas

While the technology for the transportation of natural gas from the field to the gas distribution grid in the past was limited to pipeline transfer, which required that the producing fields and consumers must be in close proximity, emerging solutions to transporting natural gas over large distances have been discovered, refined and implemented. Such innovative, technological solutions are beginning to transform the regional gas market into a truly global one, which, given time, may yet rival the global market for crude oil and petroleum products. The advantage for oil is that liquids are more easily stored, handled, and transported than gas.

The change in technological solutions has been the development of processing natural gas, itself near to pure methane, into Liquefied Natural Gas (LNG) to make it more transportable. This involves the substantial cooling of natural gas to minus 163°C, thereby transforming its state from a gas to a liquid. LNG, a colourless, odorless, nontoxic and noncorrosive liquid, and in liquid form, is 625 times smaller in volume than in its gaseous state.

The capital cost and operating costs associated with LNG have resulted in mainly wholly integrated projects and contractual arrangements that commit the gas field owner, the operator of the LNG processing plants, the shippers, and the receiving terminal facilities to long-term, typically 20–30 years, supply and purchase agreements.

The evidence that the transportation issues have been overcome—both practically and economically—is to be found that LNG is now being imported into Europe from Australia, Qatar, and Malaysia.

As LNG is a liquid when cooled, the specialist tankships are required to maintain this liquid state when crossing the oceans. The global fleet of specialist LNG tankships is estimated at around 400 ships, with a global fleet capacity average of some 164,000 m³. Capital costs of new ships are substantial, typically $ 200 million for an LNG tankship carrying 135,000 m³; ice-class ships could cost double this figure. Charter rates were thought to be of the magnitude of $30,000 per day in 2015.

There are two predominant construction designs, the Moss system (Fig. 4.3) and the Membrane system (Fig. 4.4).

The Moss system uses four or five spherical tanks, with insulation material to ensure the LNG is kept in its liquid state; the insulation is within the spheres, which themselves are supported by the ship's hull.

The Membrane system sees the insulation as a direct component of the ships' hull. Consequently, this is more optimal use of the physical volume inside the hull, and thus, the clear majority of LNG tankships now constructed use this system.

Efficiencies have also found their way into LNG tankship operation, and improvements in both vessels and infrastructure have brought loading or unloading times down to around 12 h.

FIGURE 4.3 Moss system LNG tankship. © *Society of International Gas Tanker & Terminal Operators (SITTGO), used with express permission.*

FIGURE 4.4 Membrane system LNG tankship. © *Society of International Gas Tanker & Terminal Operators (SITTGO), used with express permission.*

There are currently 28 large scale LNG import facilities in Europe, of which 24 are in the EU. These have a combined capacity of 227 billion m³, according to King and Spalding (as cited in CNBC, DiChristopher, 2018)

Transportation costs associated with LNG are nearly twice as expensive as that for gasoline and 3.5 times more expensive than crude oil, for an equivalent voyage and distance. In comparison with the capital and operating costs of the liquefaction trains and regasification terminals, these additional transportation costs, especially when amortized over a 20–30-year project life, are sustainable.

References

BP, 2018. BP Statistical Review of World Energy. London: BP.

CNBC, DiChristopher, Tom, 2018. Trump and Juncker Agree to Take Steps to Boost US LNG Exports to Europe. Available at: https://www.cnbc.com/2018/07/25/europe-will-import-more-us-natural-gas-trump-and-juncker-say.html.

5

Project management in oil and gas

Those who plan do better than those who do not plan, even though they rarely stick to their plan.
Winston Churchill

You may delay, but time will not.
Benjamin Franklin

"Spera optimum para pessimum" Latin dictum, translates as "hope for the best, prepare for the worst."

The oil industry is now synonymous with projects and project management, in much the same way as the construction and defense and aerospace industries were at the forefront of the nascent project management discipline in the post-WW2 era of regeneration. Coming later to the project management process should have meant that the oil and gas industry avoided the mistakes of the pioneers of projects, but this was not the case. Ed Merrow of Independent Project Analysis Inc. found that "... *upstream megaprojects are more fragile than their non-oil and gas industry cousins*" (Merrow, 2011).

5.1 Upstream E&P projects

Projects are usually defined as a temporary endeavor, with a definite start and target end date, with specific goals and objectives that require multidiscipline expertise and technical input that are managed by specialist project management. All projects display uniqueness, or at the very least, an element of uniqueness. A formal definition, as published by the Association for Project Management (Association for Project Management, 2004) would state: "*Projects are unique, transient endeavors undertaken to achieve a desired outcome.*"

The management of projects can be thought of in terms of the allocation of resources, the application of knowledge in a schedule time plan to achieve the specific goals in an efficient manner as possible (Badiru and Osisanya, 2013).

Once oil or gas has been discovered, the next challenge is to bring that potential into reality by exploiting the resources into, hopefully, commercially sustainable production. As with many of the oil industry processes, this is not risk-free; risk can never be eliminated, but an integral part of project management in upstream exploration and production products, and it is one of the functions of project management to reduce, mitigate or share that risk.

Projects in upstream E&P activity are characterized by huge investments with very long project lifetimes, compared to most infrastructure construction projects. The typical lifecycle for oil projects would be of the order of 20 years; for natural gas even longer by some 5–10 additional years. In addition, there will be a period of heavy capital investment with no prospect of income received from pumping oil for around 5–7 years; plenty of time and opportunity for things to go wrong.

5.2 Megaprojects

Megaprojects have become defined as those exceeding $ 1 billion. This high-cost classification obviously reflects on the complexity of the project, often in conjunction with inhospitable locations for oil and gas E&P activity, deep-water drilling, and resource-rich countries holding back easier and more accessible developments for the benefit of their grandchildren. Edward Murrow (Merrow, 2011) has conducted several longitudinal studies on industrial megaprojects and found that such mega projects are more susceptible to poor planning, interrupted project leadership, and schedule aggressiveness, often from newly appointed CEOs. Against a benchmark from other industrial sectors of a 50% success rate, only 22% of oil megaprojects were successful. The industry experience is dominated by cost overruns, time execution slippages of 30%, and enduring production difficulties in the early years after first oil or gas.

Merrow's (2011) analysis concludes that there are three factors that can explain the reason for this poor performance in oil megaprojects. The importance of Front-End Loading (FEL) is particularly vital for oil and gas projects. This is the process that combines the business case development, project scope definition, and front-end engineering design (FEED) into an executable plan and time schedule. This leads to the project proposal being put forward for a Final Investment Decision (FID).

Second, consistent project management and leadership appears more essential for oil and gas megaproject success, than in other comparable industries. Too often, the oil industry's reaction to poorly performing projects is to replace the project leader.

Finally, the adoption of aggressive and overly optimistic time schedules within the oil industry merely magnifies and accentuates the other two factors. The pursuit of the holy grail of first oil and gas by higher management by requiring project management teams to reappraise the time schedule after the project has commenced, often results in a revised project plan, with much reengineering and reprocessing, which invariably delivers first oil *after* the original milestone!

5.3 Best practices in project management

Project management skills are vital for successful E&P project execution as these projects get larger and more complex. The larger the project, the larger the risk and uncertainty. To improve performance and to minimize project delays, if not accelerate the project cycle, various commentators have promulgated a series of best practices.

- These include the establishment of a global concentration of project management to concentrate the knowledge gained to provide a repository of expertise and lessons learned, and also to bring a greater degree of standardized contractor management practice to bear.
- Developing a more sophisticated assessment of managing political risk will assist when exploring in unstable countries or those that do not have a history of political stability.
- Leverage design similarities across projects, although this seems to go against the whole characterization that projects are unique.
- Better advanced planning by the integration of contractors on a long-term basis, and to share knowledge and expertise with such contractors in a sharing of the risks and rewards that coexist in all projects.

There appears to be an objective view that speed, or reduction in time schedules, is not the answer and that oil companies need to integrate better the multivariate functional and technical disciplines present in megaprojects to prevent the sorry catalog of projects delivered late and well over budget.

5.4 Differences between project management & operations management

Operations management is concerned with the status quo, running the business as usual, with a consistent set of tasks and responsibilities. Managers who have operational responsibilities have their authority defined by the organization's structure; their responsibility limited to their own functions with a limited ability to show the flexibility of function. In traditional organizational structures, they will be working in "silos," the marketing department, human resources, finance, and so on. The operational manager's main task is optimization, and success will be measured against interim targets—annual budgets, monthly sales—these frameworks will restrict the amount of innovation required in managing or the initiative the manager can really demonstrate.

In project management, managers are responsible for overseeing change and delivering the project on time and on budget. They will only achieve this if they are exceptional *integrators* who can coordinate the project team consisting of multi-discipline workers, integrate supplier contractors, and resolve the conflicts that naturally recur with projects. There will be a constant battle between time and cost, as the ever-changing project tasks unroll over time. Project managers hold their key position only for the life of the project. Only recently has project management offered career advancement prospects, with the formation of project and program support offices. Success will be measured through the achievement of the stated goals. A project manager will live continually alongside risk and uncertainty.

5.5 Business case for projects fundamentals - payback, NPV

The oil and gas business is characterized by a high-risk, high-reward culture and needs multi-million dollar investments in order to yield significant returns for investors. As it is far easier to spend money, and not so easy to deliver sustainable long-term profits, every project or significant capital expenditure needs a sound supporting business case.

The basics of a business case are to ensure that the business benefits outweigh the costs, and by how much. Not only that, the timing of the cash flows, especially future revenue, is vital to building a robust business case. Project approvals must be assured that the financial projections are made with confidence and that all the possible costs have been identified and all the risks identified and assessed.

In purely financial terms, there are three main factors that influence investor returns and the decision to proceed. These factors ensure investors are protected from the effects of inflation, the premium attaching to the additional risk above what is perceived as "normal" and the opportunity cost, or the return that could have been earned on the next best investment.

Therefore, a business case will be supported and largely influenced by the quantitative business economic analysis, which can take many forms, such as Return on Investment, Payback Period, Net Present Value, and Internal Rate of Return.

The reader is directed toward any specialist management accounting textbook to learn in detail of these methods, their advantages and drawbacks; one such is listed in the Further Reading section at the end of this Chapter.

Payback Period (sometimes referred to as Payout Period) is defined as the length of time for the initial investment to be repaid out of the net cash inflows from the project. Organizations usually set a "hurdle" rate if using this method of financial appraisal, and acceptable projects must exceed this hurdle rate, i.e., show a Payback Period *shorter* than the hurdle set by the business. With competing projects, all of which meet the hurdle criteria, then the project with the shortest Payback Period would be chosen in preference.

Payback Period favors short investments, but completely ignores the net cash inflows after the payback period has been met; these additional cash flows occurring after the payback period could be considerable and generate large benefits in later years. By ignoring the total life cycle cash flow, the Payback Period is never used by major oil and gas companies as the sole determinant of business case probity.

The other significant drawback of the Payback Period as a financial business appraisal method is that it completely ignores the time cost of money. Due to inflation, the value of money loses its value, such that a dollar today will not buy as many goods in (say) 10 years' time. As it would be improper to rate cash flows in an oil project over a 20-year period as having the same purchasing power or value, the industry uses the **Net Present Value (NPV)** method of appraising projects and their net benefits.

In the Net Present Value method, future cash flows are discounted at a given or assumed discount rate to arrive at the present value of each future cash flow. By adding these and subtracting the capital expenditure required at the start of the project, a Net Present Value can be derived.

If this Net Present Value is positive, then the project can be approved, as the net cash benefits have already accounted for inflation, risk, and opportunity cost within the assumed discount rate. Conversely, projects with negative Net Present Values should be rejected, as the negative value indicates that the project will be a cash drain on the organization. In the case of competing projects, the optimal choice is that project with the highest (and positive) Net Present Value.

The NPV is very much the go-to choice of business financial appraisals as it takes account of the timing of future cash flows and uses all the cash flows from the entire life cycle of the project. In addition, projects with a positive NPV are increasing the wealth of the organization, an objective that both aligns with the aims of the business and shareholders wishing to increase their wealth in their holding of the company.

A similar methodology is the Internal Rate of Return (IRR) which can be defined as that discount rate that, when applied to the project's future cash flows, produces an NPV of exactly zero. It can, therefore, be said to represent the yield of the project investment opportunity.

5.6 Risk and uncertainty

All businesses face risks of some nature, and the oil and gas industry, with its high-value projects spanning many years, is no exception. The industry, therefore, deals with risks in a highly professional manner. Not all risks impact negatively on a project; for instance, the risk on the recovery rate of an oil reservoir may turn out to be a positive one, if the planned theoretical rate of recovery is surpassed. The key is finding the balance between risk and reward, not being completely risk-averse and avoiding sensible, acceptable risks, and not being unnecessarily reckless and ignoring real risks that could jeopardize the future of the business.

Risk is anything that interrupts or jeopardizes the organization from achieving its strategic objectives or a project from reaching its stated goals. In the oil and gas sector, timescales associated with projects are long in comparison with many other industries, therefore there is more time for things to go wrong. If things do go wrong, the size of oil investments will usually mean significant and lasting problems.

Uncertainty should not be confused with risk; it is present in all aspects of oil's value chain, from wellhead to petrol pump. Uncertainty can be explained by an uncertain outcome of events with an unknown probability of occurrence. Uncertain events occur without anyone in control and can never be measured or quantified. When you do not know the outcome of any activity, then you are uncertain.

BOX 5.1 Risk management

Risk management is the state of having a contingency ready to respond to the impact (good or bad) of occurrence of the risk, such that risk mitigation or risk exploitation becomes an intrinsic part of the project plan.

Badiru, A.B., Osisanya, S.O., 2013. Project Management for the Oil and Gas Industry: A World System Approach, USA: CRC Press, Boca Raton, Florida.

Both risk and uncertainty talk about future losses or hazards; whereas risk can be quantified and measured, there is no known way of ascertaining uncertainty. Risk can be mitigated by good project management, but even the best management cannot remove uncertainty altogether (Box 5.1).

5.7 Network planning

In the prosecution of projects, oil and gas companies must carefully manage the three constraints known as the Iron Triangle - time, cost, and quality.

Network analysis is a technique used by project managers to plan and allocate resources to minimize total project duration and project costs. Critical Path Analysis is a complementary technique to identify tasks or activities, which must be executed on time and in a correct and logical order, for the project to be completed on schedule.

There are five basic steps in this network analysis methodology:

- Identify all the activities that need to be undertaken as part of the project
- Estimate the time it will take to complete each activity (see Box 5.2)
- Identify the logical order in which the activities need to be completed
- Construct a network diagram
- Calculate the total project duration

BOX 5.2 Problems with task estimation

Underestimation of projects' complexity and risk.
Underestimating the cost to attract and maintain the labor force.
Underestimation of productivity loss associated with climatic working conditions and shortage of daylight hours in northern regions.
Shortages of skilled labor and competition between employers seeking to attract labor in a tight labor market.
Lower-than-expected productivity, due to mismanagement of construction.
High labor turnover due to the harsh working environment.

BOX 5.3 Problems with managing the project

Inadequate field experience in engineering and procurement personnel.
Late delivery of engineering, equipment, and materials.
Poor project controls.
Inadequate plan of execution; poorly defined tasks and division of responsibility.
Lack of knowledgeable leadership in engineering, procurement, construction, and start-up.
Inexperienced project management personnel.
Lack of standardization.
Poor communication, teamwork, and alignment of interests between contractors.
Poor site organization, excessive time wastage, and loss of productivity.
Joint venture partners and contractors not perfectly aligned.

From the network diagram, it will be possible to identify the critical path, on which the activities which, if delayed, will delay the completion of the project. A project's critical path is understood to mean that sequence of the critical (or slowest) activities that connect the project's start activity to the finish activity.

The objectives of project planning are to optimize the project process to prevent problems in the process, especially those associated with escalating costs. This can be achieved through a systematic evaluation of the project's constituent activities with their duration and logical linkage; problems with managing projects are shown in Box 5.3.

5.8 Contracting options – EPC, EPCM, BOOT, DB, Turnkey

For succeeding in its project management objectives, the oil and gas industry is dependent on a vast network of contractors, partners, and suppliers; managing these relationships is critically important. Oil and gas organizations must explore the balance between in-house expertise and buying in services from the open market, the "make or buy" decision.

Finding the optimal approach to a project is critically important. Clearly, there are advantages and disadvantages for each option, and careful analysis of all the options is necessary to arrive at the optional solution or approach.

Engineering, Procurement and Construction [EPC] as a contracting option, provides for a single point of reference; a guaranteed completion date, price, and performance; a clear division of obligations and liabilities; supplier-assumed risks leads to a higher price. There is full dependence upon one contractor to carry out detailed engineering design, procure equipment and materials, and then construct to deliver a functioning facility or asset.

Engineering, Procurement, Construction and Commissioning [EPCC] is the same as EPC with the addition that the contractor commissions and tests all facilities prior to commercial start-up.

Engineering, Procurement, Construction and Management [EPCM] is where the contractor is effectively an assistant to the project's owner; the overall risk and control sit with owner; there is a complex contract build up and risk allocation. The contractor is not directly involved in construction but responsible for administering the construction contracts.

Generally used by Governments to develop public infrastructure, Build-Own-Operate-Transfer [BOOT] involves the private sector to finance, design, operate, and manage the facility developed for a fixed time period. This facility then transferred to Government for no additional cost. Examples of Build-Own-Operate-Transfer could be airports, bridges, motorways/toll roads, and tunnels.

Using a Design-Build [DB] contracting system, the project's owner contracts a single firm to design and build the project. The contractor's appointment is based on an outline design/brief to understand the project owner's intent. This contracting approach needs a clearly defined scope of work before signing the contract. It is essential that the project definition is fully understood by the contractor to avoid conflicts. This method often used to shorten time required to complete project. The contract is awarded before starting any design/construction, so a cost-plus contract is usually awarded. The owner's participation is extensive during the whole life-cycle of project—monitoring of costs, quality, schedules—usually achieved by owner with a team of qualified professionals. DB contracts are used for relatively straightforward projects, where no significant risk or change is anticipated, and where the owner can precisely specify what is required.

A Turnkey contract is one where the Turnkey Contractor, normally the drilling contractor in E&P, agrees to drill a well to an agreed specification for a fixed price. If the Turnkey Contractor fails to deliver the well to the agreed specification, no payment is due to the Contractor. Basically, an arrangement under which a contractor designs and constructs a project, building, or a plant, for sale when completely ready for occupancy or operation by the client or customer.

5.9 Fossil fuels future development

5.9.1 Shale oil

History will determine whether the explosion of shale oil (and gas) in the USA in the 2010s will be viewed as a second seismic shift, after the 1960 formation of OPEC, in oil's evolutionary journey since the Pennsylvania oil boom in the 1880s. Even as late as 2011, the US Bureau of Land Management was saying:

> *There are no economically viable ways yet known to extract and process shale oil for*
> *commercial purposes*
> ***Herkenhoff, 2014***

Fueled by America's consumption-driven economy, the quest for cheap energy, and a determination to energy self-sufficiency, by 2013, the USA had overtaken Saudi Arabia to become the world's largest producer of oil. Encouraged by better economics on the back of rising crude oil prices making shale extraction increasingly viable, and mineral rights vested in the land-owner, and not the state, Midwest shale production has made a mockery of this prediction. Falling oil prices during early 2015, after OPEC took the decision in November 2014 to maintain production in the face of declining demand, ostensibly to "break" the US shale oil drillers, took its toll. But shale oil production can be shut-in and then ramped up to cope with fluctuating demand far easier than conventional oil fields.

Developing technology, such as hydraulic fracturing (Fracking) to artificially induce the cracking of the shale rock sediment trapping the oil to then release this, and horizontal drilling techniques that allow for a greater shale contact area per borehole have helped, and turned potential resource plays into a reality that is currently benefitting the US economy with much cheaper energy prices for gas and oil than its major competitors, such as Europe. The stark facts are that the US oil production levels have doubled in the last decade, and US gas production has risen by 40% in the same period (BP, 2018).

OPEC has failed to bankrupt the US shale producers as they had set out to do, and what was being planned and built as import terminals for LNG are now being reverse-engineered into storage and export facilities to export shale-produced gas and oil products.

Where the mineral rights are owned by governments, then opposition to fracking and its associated noise, and polluting effects have had a focused target, and to date, this has prevented large-scale commercial fracking starting in the UK, France, and Germany. Whether this opposition will ever be overcome in order to provide greater national energy security remains to be seen; the economic barriers against fracking have been largely broken, due to the USA's efforts to embrace and improve fracking technology in order to displace Arab-sourced crude oil and LNG.

5.9.2 Unconventional crude oils

Aside from shale oil and gas, there are two other major sources of what are known as unconventional crude oil sources; these are oil (or tar) sands and coalbed Methane. The term unconventional is used to denote a hydrocarbon source that must undergo a process before running in a conventional refinery as a conventional crude oil with gravity in the range 10 −50 degrees API. Obviously, any such additional process puts these unconventional sources at a cost disadvantage.

Oil sands represent the largest source category of unconventional crudes, with the sands in the Athabasca basin of Canada and the Orinoco oil belt of Venezuela dominating this sector. Oil sands are effectively extraheavy crudes which contain a high level of impurities, chemicals, and especially bitumen. Development costs are naturally very high, so oil prices need to be trading consistently high—probably in excess of

$60/bbl—to make extraction a sustainable and profitable operation. Production levels showed an output of 2.53 mm bpd in 2015, representing 62% of all Canada's crude oil production (Canadian Energy Research Institute, 2017).

National Resources Canada (2017) states that Canada has a total of proven oil reserves of some 171 billion barrels, making it the world's third-biggest reserves behind Venezuela and Saudi Arabia; some 166 mm bbls of this figure are located in the Alberta tar sands.

Although the exploration risk and cost is low, as the tar sands are readily visible, extraction is a costly, dirty, and environmentally unfriendly process. Tar sands on or near to the earth's surface are strip-mined and then treated with very hot water to separate the petroleum elements from sand; once this is completed, then the bitumen-heavy sand is further treated to separate out the bitumen itself and then this either has to be upgraded to a synthetic type of crude oil or blended with lighter products to facilitate transportation to a refinery.

The bitumen from oil sands is a very heavy form of crude and sells at a vast discount to conventional crudes such as WTI. Crude oil production from oil sands overtook conventional Canadian crude oil production in 2010, and in 2013 was stated to be 1.9 mm bpd, compared to 1.5 mm bpd for conventional sources (Natural Resources Canada, 2014).

If more natural gas penetrates the transportation market over the near future, then traded natural gas and LNG prices will increase to the point of convergence with oil price levels. This has an impact on the costs of oil sands production, as the natural gas used to generate the heat required for the extraction process becomes too costly, and therefore, tar sands production costs become unsustainable.

5.9.3   Biofuels/fossil fuels mix

The term biofuel refers to a fuel that is produced by synthesizing a living organism by one of several conversion processes. In liquid form, these biofuels can be mixed with fossil fuels to reduce the carbon basis of traditional fossil fuels. Very often, governments mandate the inclusion of a biofuel element to help drive global economies toward a low-carbon future.

In Europe, legislation already exists to force producers and wholesalers to include 5% biomass content by weight in conventional petrol and diesel. This level of bio/fossil mix has been accepted by the manufacturers of petrol and diesel engines, such that engine warranties are not invalidated. The consumer is largely unaware of this biomass element – there is no special product marking on pumps, and the existing vehicle parc operates on such blends perfectly normally. Increasing the biomass blend share to 10% by weight would impose further challenges, as some older cars and their engine management system would not be able to operate normally, and thus, a new E10 petrol blend (10% ethanol in 90% fossil fuel gasoline) could require a discrete supply and distribution channel, even requiring separate dispensing pumps at service stations.

The supreme advantage of any biomass/fossil fuel mix is that the existing bulk liquid fuel distribution system would work without modification and not need significant capital investment.

5.9.4   Synthetic production

Synthetic fuels are costly to produce and to convert in any transformation process into useable consumer products, especially those designed for application in the transportation sector. While synthetic fuel technology has been used commercially since the 1940s when Germany was using the Fisher-Tropsch process that allowed the conversion of plentiful coal deposits into gasoline, the high costs of the plant, operation and maintenance are further compromised that the end product, as an alternative for conventional gasoline, has a commodity-value that is both volatile and unpredictable.

The exploitation of stranded gas assets overcome some of the economic dis-benefits, as in this case, the alternative costs of pipelining this gas over long distances to consuming cities determine the opportunity cost of synthetic gasoline to show an economic desirability. South Africa's Sasol is possibly the world's biggest producer of synthetic gasoline. This has resulted from a combination of political isolation during the apartheid years and the inability of South Africa to import gasoline and the geological fortune of significantly large indigenous coal deposits.

Other examples of synthetic fuel production are found in Malaysia, where a Shell plant processes natural gas into low-sulfur diesel and waxes, and in Finland, where UPM—a paper pulp manufacturer—produces biodiesel from the waste from paper manufacture.

5.9.5   Gas-to-liquid production

The Gas-to-Liquid (GTL) process converts natural gas to diesel fuel, commonly using the Fisher-Tropsch process. This can be thought of as a reverse of the refinery chemical decomposition process. Unlike a refinery, where long hydrocarbon chains present in crude oil, are broken down into shorter chains to form more useful petroleum products used for transportation fuels, the Fisher-Tropsch process converts natural gas feedstock into longer chain hydrocarbons.

The Fisher-Tropsch process, developed by chemists in Germany in the 1920s, begins with partial oxidisation of the methane in natural gas. This produces carbon monoxide, hydrogen and water, and carbon dioxide, and the excess of the latter is removed using aqueous solutions. By removing the water, synthesis gas is produced, which, when allowed to react over a catalyst, will produce liquid hydrocarbons, which can then be cracked to form diesel fuel.

The rationale behind this GTL technology is that liquids are far more easily stored and transported than a gas. The world's largest GTL plant is Shell's Pearl, situated in Qatar. New technological solutions are being developed for GTL, as the Fisher-Tropsch becomes an effective choice for the conversion of unconventional, hard-to-reach gas that might otherwise be left underutilized as a stranded asset.

5.9.6   OPEC's waning power?

At their November 2014 meeting, OPEC decided to maintain production in the face of stalled global demand and increased US shale oil production volumes. This strategy, which defied all classical economic knowledge of decreasing production when demand weakens in order to maintain price, was largely at the behest of Saudi Arabia and was designed to choke off the US shale producers.

This strategy backfired spectacularly as oil prices plunged to record lows during the first half of 2015, events which also conspired to hurt those OPEC producer countries that have more structural domestic economic issues, such as Nigeria and Venezuela. Undoubtedly, the price falls did cause some small scale shale oil producers in the US to shut up shop but more decided to mothball or merge to ride out the low price storm.

In November 2016, OEC decided upon a 6-moth production cut to shore up prices, and then bring alongside non-OPEC production cuts in December 2017, including a Russian promise to cut some 500,000 bpd of oil production in a last bid to stabilize and give support to price levels in the oil markets.

Perversely, by cutting back on production that naturally means that prices must rise to satisfy demand if that is assumed to stay static, is a consequence that also benefits US shale producers, who need higher prices to compensate them for higher production costs than the traditional oil field owners in the Middle East.

OPEC has limited means at its disposal now to influence the market. Saudi Arabia, for long the world's "swing" oil producer able to either flood or restrict the market, no longer holds that sway. US shale producers are adept at turning on and off the taps and can do it far quicker than conventional oil field drillers. Continued efforts to reduce production by OPEC will be compensated by the USA ramping up production, motivated by President Trump's desire to reduce energy dependence on Arab oil. OPEC, and especially Saudi Arabia, is now caught in the crossfire within a battle they started.

Despite the rhetoric emanating from the Middle East that non-OPEC producers are "free-riding" on the more buoyant prices engendered by OPEC policy of production cuts, it is now hard to see how OPEC can wrest back control of, and influence over, the global free market for oil. Increasing prices will reduce oil revenues for many OPEC country economies that are overreliant on the petro-dollars to fund their fiscal budgets; there is no longer any certainty that US shale producers will be forever disadvantaged, they have already shown remarkable resilience and will be better placed the next time around.

OPEC now controls just one-third of the world's oil output, its smallest share for almost 30 years. Increasingly the cohesion that has held OPEC together since its formation in 1960 is showing signs of strain. Only Saudi Arabia now has the spare production capacity to influence the market; some OPEC countries, such as Nigeria, are exempt from the production cutting strategy; will the Saudi-Russian concord on production restrictions stand a test of time? The balance of power has inexorably shifted, and the world can no longer be held hostage to oil supply restrictions and high prices as it once was in 1973.

5.9.7 Sustainable development

Sustainable development means that already available resources are managed in a manner conducive to ensuring wealth creation on a long-term basis. This is, therefore, not to be confused with whether oil and gas are sustainable alongside renewable forms of energy. Sustainable development is thus a link between society's needs, global, and national economies, and protection of the environment.

The challenges in balancing all these three disparate interests will tax the oil and gas industry. To encourage the economic growth in developing countries, conventional fossil fuels will be required in plentiful supply at reasonable costs until renewable fuels can provide a credible alternative energy source.

Therefore, the oil and gas industry has a vital role to play in making energy available while at the same time protecting the environment. So sustainable development for oil and gas should mean utilizing the known available resources for the benefit of society, but not on the assumption, that means indefinite supply. This different approach will require a change in strategic direction, and major oil and gas companies will expand their organizations by investing in whatever alternative fuels become available. They will also have to make effective use of the limited and depleting fossil resources at their disposal, such that petroleum products provide a continuum to ensuring more of the world's population enjoys and benefits from the quality of life experienced and enjoyed in the developed western world.

Sustainable development will be assured if the major oil and gas companies set a precedent and high ethical standards that minimize any impact of their operations on the environment and will increasingly work with all parts of society and embrace the social objectives of the communities in which oil and gas operations exist. If we assume that peak oil will eventually approach, the value of what will be increasingly precious fossil fuel resources must be husbanded and used in an optimized manner to support developing countries and societies. New approaches to extending the life of existing resources and new energy sources must be seen as a complementary adjunct to the traditional oil and gas and not as a competitor.

5.9.8 Meeting demands of global population growth

The United Nations projects that the world's population will reach 10 billion people in the year 2056; the last UN report on global population for 2015 was 7.4 billion. (United Nations, 2017). Most of this increase will emanate from developing nations and will be characterized by an explosion in the industrialization of those nations, with many people also moving from rural habitats to live in cities and towns. This will, of course, increase energy demand.

Fossil fuels will remain a vital component of the future energy mix, despite environmental concerns in the developed world, which will shift primary energy consumption away from carbon into solar, wind, and other renewable sources. Electric cars will replace conventionally fueled automobiles and start to reduce carbon dioxide emissions on a significant scale from 2035 onward.

However, the population increase forecasted in developing nations will require freedom of labor movement to sustain the predicted industrialization, and this will require energy. Oil and gas will empower this industrialization, for the simple reason that there is nowhere near a suitable, or as easily transportable fuel that provides so much energy content per liter or per tonne.

It is inconceivable that US oil demand will dip much below current levels of 19 million bpd, or that Europe, where environmental concerns are most attuned, will see a significant drop in demand, for as health care and treatments improve, life expectancy increases, and with it, people living longer maintain their lifestyle, including consumption of energy. Many commentators, including OPEC, are predicting that oil demand will grow by some 10% during the next two decades to a figure of more than 104 million bpd by 2040.

Demand for oil will remain inelastic with regard to price, and demand will continue to be fueled by the global population increases, economic growth, and disposable income. So, the "peak oil" theorists' arguments may yet stay; the more significant question is whether international and national oil companies' single biggest challenge can be met? Can this burgeoning demand be fulfilled by finding even more sources of fossil oil and gas? If so, the exploration and production divisions will assume a prime position within this next intriguing phase of the oil and gas industry.

5.9.9   Future transition from oil and gas to electricity

At first sight, it is hard to understand a transition from oil to electricity, as these are fundamentally completely different forms of energy. In addition, given oil's dominant share of the primary energy market (Hydropower accounted for a 6.8% share in 2018; renewables 4.0% share, according to BP's Statistical Review of World Energy, 2019) any transition pathway has been almost impossible to define or to assign a feasible and pragmatic time-line.

But, the world and its people are changing. The growth of electricity is far greater than that of oil, and it is predicted that electricity will overtake oil as a primary energy source by 2040. Costs of electricity generation are also falling, removing some of the barriers against fuel substitution. There appear to be three irresistible factors that will drive the transition.

First, the growing availability of electric vehicles (EVs) will begin to make significant inroads into the small and medium-size car market, when the technology developed for the high-end market (such as Tesla and Porsche) is harnessed and refined by the manufacturers for incorporation into their mainstream popular models. As more and more EVs populate the global car parc, the demand for fossil-fuel-derived transportation gasoline and diesel will decline. Increased range of battery driving—mileage will help, as will the reducing capital cost of acquisition. EV growth may yet be tempered by a lack of recharging infrastructure dampening demand; converting existing petrol service stations would be an obvious remedy, but only if battery recharging times can be lowered significantly.

Second, pressure from activists and consumers alike on governments to do more about limiting carbon emissions will only serve as a continual motivation in diminishing reliance on fossil fuel energy forms.

Third, financial institutions such as banks and insurers, are now reducing their risk exposure to fossil fuel investments, due to climate change pressures from shareholders and active protestors.

A report published by the Grantham Institute at Imperial College London (Expected the Unexpected, February 2017) suggests that Solar Photovoltaics (PV) could supply 23% of generated power in the world in 2040, rising to 29% by 2050. In this projection, coal would have been eliminated as a power source, and natural gas limited to a paltry 1% market share of generation.

In the same report, the prediction is that EVs would have captured 35% of the road transport market by 2035, and by 2050, two-thirds of the market. If this extrapolation holds true, then some 25 million bpd of oil demand will be replaced by EVs in 2050.

Natural gas distribution infrastructure—commonly referred to as a "grid"—was often funded by governments through their nationalized gas companies. In the developed world, the consumer gas market has been privatized, and with no political or social agenda, these private companies have become marketing-focused suppliers. Despite continued housing development in rural areas in both developed and developing countries, expansion of the gas grid to fully interconnect all new-build housing is far from guaranteed, unless the development is adjacent to the existing grid.

Homes outside the reach of any gas-supplying grid have, in the past, turned to kerosene or gasoil fueled boilers to provide heating and hot water; these fuels being more controllable than electric storage radiators or panel heaters. In such cases, the transition to rural electricity would appear relatively untroubled, as all homes will already have their electricity supply, thereby facilitating a switch to heating and hot water generated either by solar panels or electric systems.

It is also informative to note that several oil supermajors are investing heavily in EV battery producers and charging networks. Having reaped the benefits of a fully controllable and vertically integrated supply chain—effectively from well-head to petrol pump—it is not surprising that this successful business model is applied in one of the fastest growing energy sectors.

Total, the French oil company, is rapidly becoming a utility provider with investments in local electricity retailing, power generation and distribution, and battery manufacturing. Total share a belief in the assumption that electricity will be the primary energy of the 21st Century. Shell has plans to become the biggest player in electricity generation by 2030 (Oilprice.com, 2019) by investing $2 billion into cleaner power generation each year.

Big Oil has a desire to become Big Power, and this is an ambitious and realistic possibility. The existing large power generators will find it hard to resist the financial muscle or corporate will of the oil supermajors. The move into dominating the power utility sector by the supermajors will *"create a new breed of gigantic energy-controlling monopolies"* (S&P Global Platts Insight, 2019).

5.10 Conclusion

Oil and gas are not dead, or even not dead yet! Despite politicians' desire for a no-carbon future, it is just not practical even in a highly developed, socially inclusive country with a robust economy. The costs of transitioning to a low-carbon future and the required infrastructure, the economic reluctance to writing down productive assets, such as motor cars and train locomotives, the loss of taxation income on transportation fuels are the practical issues that the theoretical governmental mandarins have not thought through.

In the developing world, bulk liquid fuels containing high energy content are easy to transport, store, contain, and use efficient and proven technology and proved mass transportation, essential to the movement of labor driving economic growth and increased affluence.

So, carbon rich, liquid fuels will have an important part to play in society for years to come. This is not saying there will not be change, but the oil and gas industry have never been afraid to face up to the challenges of change and have a rich history of coming through those challenges since the late nineteenth century.

When I first joined Texaco in 1973, my immediate manager told me that:

> *… if cars in the future were to be powered by pills, then Texaco would be involved in the manufacture, distribution and marketing of* those pills to motorists.

In those days, climate change was a long way off, a low-carbon society unheard of, but oil companies are nothing if not adaptable, and it is very likely that Exxon, Shell, and BP will still be household brand names trading in another 40 years' time. The oil industry is far from dead.

References

Association for Project Management, (2004). www.apm.org.uk. (Online) Available at: https://www.apm.org.uk/body-of-knowledge/context/governance/project-management/ (Accessed 11 November 2018).

Badiru, A.B., Osisanya, S.O., 2013. Project Management for the Oil and Gas Industry: A World System Approach. CRC Press, Boca Raton, Florida, USA.

BP, 2018. BP Statistical Review of World Energy. BP, London.

Canadian Energy Research Institute, 2017. Canadian Oil Sands Supply Costs Abnd Development Projects (2016–2036. CERI, Calgary, Alberta.

Grantham Institute at Imperial College London, 2017. Expect the Unexpected (Online).

Herkenhoff, L., 2014. A Profile of the Oil and Gas Industry. Business Expert Press, LLC, New York.

Merrow, E.W., 2011. Oil and Gas Industry Megaprojects: Our Recent Track Record. Society of Petroleum Engineers, Houston, Texas, USA.

Natural Resources Canada, 2014. Energy Markets Fact Book 2014–2015, s.L. National resources Canada.

Natural Resources Canada, 2017. Oil Resources (Online). Available at: www.nrcan.gc.ca/energy/oil-sands/18085 (Accessed 12 October 2018).

Oilprice.com, 2019. How Big Oil Could Become Big Electricity (Online). Available at: https://oilprice.com/Energy/Energy-General/How-Big-Oil-Could-Become-Big-Electricity.html (Accessed 9 October 2019).

S&P Global Platts Insight, 2019. Big Oil's Electric Dreams Could Create New Energy Cartels, by Andrew Critchlow, (Online). Available at: https://blogs.platts.com/2019/03/20/big-oil-companies-electric-dreams-energy/.

United Nations, 2017. World Population Prospects: The 2017 Revision (Online). Available at: https://population.un.org/wpp/Download/Standard/Population/ (Accessed 21 October 2018).

Further reading

Atrill, P., McLaney, E., 2016. Financial Accounting for Decision Makers, eighth ed. Pearson Education Limited, Harlow, UK.

Energy Information Administration, (2018). eia.gov. (Online) (Accessed 2 November 2018).

Energy Institute & Joint Inspection Group, 2013. Quality Assurance Requirements for the Manufacture, Storage and Distribution of Aviation Fuels to Airports, first ed. Energy Institute, London.

HM Government, 2018. Digest of UK Energy Statistics (DUKES) (Online). Available at: https://www.gov.uk/government/collections/digest-of-uk-energy-statistics-dukes (Accessed 21 October 2018).

James, T., 2008. Energy Markets: Price Risk Management and Trading. John Wiley & Sones (Asia) Pte. Ltd, Singapore.

OPEC, 2018. OPEC Web Site (Online). Available at: https://www.opec.org/opec_web/en/about_us/23.htm (Accessed 28 October 2018).

Project Management Institute, 2018. Project Management Body of Knowledge (Online). Available at: https://www.pmi.org/pmbok-guide-standards (Accessed 11 November 2018).

6

Distributed generation systems (DGS)

The objective of any off-grid system is to benefit the local community in various ways, but especially so for the poorer communities in developing countries. These benefits may include water pumping used within a village and/or farming environment, the generation of electricity for lighting, heating, and cooking, agricultural processing tasks, e.g., winnowing, threshing, grinding, and for getting rid of saline water (pumping). These are just few examples, as there are many other applications, but in particular, there is the need for a stable and regular supply of electricity to help the local population, including businesses and industries.

6.1 DG key factors

The following points are some of the main benefits DG (off-grid) system (technology project):

DG systems can provide a lower cost than a system connected to the main grid; in the long term, i.e., each household will be self-sufficient in energy.

Using RE sources for off-grid systems means a reduction in CO_2 and other greenhouse gases.

With the correct installation and insulation, off-grid systems can provide better energy efficiency.

6.2 DG general applications

The large number of applications, which DG systems can provide, whether in the form of energy supplied for rural applications or urban areas, the main principle is always the same, energy self-sufficiency. The main points of DG applications can, therefore, be summarized in the following points:

1. Self-Sufficiency, i.e., meeting energy need for the individual and local community
2. Cogeneration: Capturing waste heat for efficiency and for energy enhancement
3. Supplementary power: For the main grid (feedback - if connected to the main grid)
4. Metering: Income for selling excess electricity generated back to the main gird (if connected to the main grid)

Rural Electrification. https://doi.org/10.1016/B978-0-12-822403-8.00006-0

5. Support for the main grid: Reduce congestion during peak demand (if connected to the main grid)

In general, the DG technologies' size range differs in accordance with the type of model being used and the type of energy source for which the device is designed. However, the range for nonrenewable sources models of DG falls between 30 kW and 15 MW. The size range of the following models has been estimated as follows:

Micro-turbines	30–200kW and can be up to 1,000 kW
Mini-turbines	200–1,000 kW
Small turbines	1,000–15,000 kW
Industrial turbines	1–15 MW
Reciprocating engines	30–15,000+ kW and 60 MW
Fuel cells	30–1,000 kW
Recent development may range from	5–1,000+ kW
Fuel cells hybrid	200–1,000+ kW

6.3 DG barriers

It is possible to reduce DG barriers by investigating three main important aspects. These three main points can be applied to DG, regardless of whether they are connected to the main grid or not:

1. Technical Barriers
2. Business Practice Barriers
3. Regulatory Barriers

The technical barriers are for the need for uniform standards, interconnection, and power control.

The business practice barriers are for the need for adopting the standard commercial practice, business terms for interconnection, and tools to determine the value and impact of distributed power at any point on the grid.

The regulatory barriers are associated with tariffs and utility incentives to match with DG, conditions/terms to interconnect, regulations compatible with DG for competitive, utility market, and the establishment of dispute resolution for DG projects.

Finally, introducing DG can be a threat to the current business model, as an increase of DG penetration means a decrease in revenue for existing main power stations and that in itself is a barrier for the expansion and/or the development of DG network systems (Boxes 6.1 and 6.2).

6.4 Extension of the existing grid (EEG)

6.4.1 Introduction

Most of the electricity supplied from the main commercial power stations have most likely originated from fossil fuels (or nuclear-generated). Possible electricity grid

---

BOX 6.1 Distributed energy

A joint DTI/Ofgem Review of Distributed Generation was published alongside the EWP. The Review identified a number of barriers to DG, including cost, a lack of reliable information, electricity industry issues, and regulatory barriers. One of the main conclusions of the Review was that the UK energy regulatory regime *"was established to meet the needs of large centralised generation and aspects of the system disadvantage smaller distributed generators."*

Ofgem, 2013. Distributed Energy - Current Issues.

BOX 6.2 Costs associated with implementing and maintaining a DG system

---

The financial reality of the adoption of a DG system is that it is generally reliant on the commercial viability or short payback period of the project. Although strong social incentives such as reduction of GHG emissions and the positive connotations that installing a DG system will have on an organization are relevant, the absence of a reasonable payback period or proven commercial viability is likely to result in an unsuccessful DG installation attempt.

Sustainability Victoria, 2011. Distribution Generation Experiences Analysis – Survey Report 2010. http://www. sustainability.vic.gov.au/resources/documents/distributed_generation_experience_analysis.pdf 19/04/2013.

---

extensions in many developing countries may not be an option, although efficient at the point of use, extending the electricity grid can be very costly in terms of capital expenditure, as well as regular maintenance (Fig. 6.1).

As the rural systems tend to be used during peak hours and the capacity usage is low, the system losses tend to be higher compared to the electricity supplied from DG installations. The benefits (Fig. 6.1), however, can be found when there is a possibility of a DG system being connected to the main grid. The control mechanism between the main grid and the DG system can play a major role regarding continuous electricity supply, better distribution of electricity demands, and consequently, an overall cost reduction.

6.4.2   Grid extension to the countryside

The World Bank has put forward a number of factors that should be considered prior to any extension of the main grid into the countryside. The community living in a rural area should first have some basic infrastructure, such as water supply and roads. In addition, there should be sustainable growth in farming (or a positive agricultural situation), as well as market access to these farming products. Furthermore, there should be an already existing development activity in the village with possible future improvement in the living standard, plus expected growth in the local businesses.

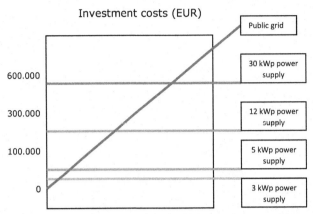

FIGURE 6.1 Illustration of the cost for grid extension in relation to distance. *Redrawn and edited from ARE (Alliance for Rural Electrification), Undated. Hybrid Power Systems Based on Renewable Energies: A Suitable and Cost-Competitive Solution for Rural Electrification. http://www.ruralelec.org/fileadmin/DATA/Documents/06_Publications/Position_papers/ARE-WG_Technological_Solutions_-_Brochure_Hybrid_Systems.pdf, 14.06.2010.*

According to Holland et al. (Undated) providing electricity, e.g., for a village in Thailand, the World Bank has been following priorities factors such as:

1. The size of the village
2. Access to good roads
3. Distance to the distribution network and/or villages nearby already being supplied with electricity
4. Approximate number of consumers
5. Approximate number of established businesses
6. Approximate number of rural industries
7. Comparison of other similar work done in other countries to the new location
8. Availability of public facilities.

The main grid can also be used in conjunction with a number of DG connected together, i.e., DG network, where a number of villages close to each other are located at a viable distance from the main grid. This kind of network, where both the main grid and a number of DG are connected, should be able to facilitate a stable electricity supply under the umbrella of national regulations. It will also provide long-term grid interconnectivity, bigger market orientation, and rural business and industrial growth.

6.4.3   Summary (EEG)

1. To provide a continuous supply of electricity on a 24-hour basis
2. Unlimited loads
3. Support productive loads
4. Positive aspects of load growth
5. High initial cost for rural and remote areas
6. Uneconomical for small loads

7. Grid extension to a village can benefit all the community regardless of whether some households can afford to pay for it or not (e.g., street lighting, health, and education services)
8. Can encourage economic development in the rural areas
9. Can start and establish socioeconomic development within a village and beyond
10. It may reduce government subsidies for the rural areas when the grid is connected to an urban area, i.e., a higher standard of living in the urban area will contribute financially to the cost of the electrification of the rural areas
11. Low cost of electricity usually happens when power stations are able to provide a large number of customers, including rural areas
12. Governmental pollution control will be applied to the countryside as well when extending the grid to rural areas, especially when the power station uses fossil fuel to generate their electricity, i.e., environmental monitoring for the countryside

6.5 Off-grid systems

As have been mentioned at the opening of this chapter, off-grid systems in a developing country can benefit the village community in a number of ways, such as:

1. Water pumping used within a village and/or farming environment
2. Generation of electricity for lighting, heating, and cooking
3. For agricultural processing tasks, winnowing, threshing, grinding and for getting rid of saline water (pumping).

There are, of course, other applications that can be considered as well, particularly when there is a need for stable and regular energy supply to help local businesses and their local industries. Higher power of electricity output will be needed in situations like these, and consequently, the cost can be higher than the cost of the applications mentioned above.

Various technologies can be used for decentralized operations. Tables 6.1 and 6.2 provides some size ranges for such technologies. Tables 6.2 and 6.3 provide a comparison between the prices of 2007 and 2016 for some of the renewable energy technologies (DG) and the prediction that in another 10 years, by 2030, the prices will go down further, which may make it even cheaper than the usage of fossil fuels.

6.6 Important factors

From a large number of DG research project sources, the following points have been selected and summarized as the key factors that should be considered before and during the work related to any DG technology project in a rural area:

1. Present and future level of energy need for the community
2. Procurement cost and cost of maintenance

Table 6.1   Summary of DG benefits and services.

	Energy cost saving	Saving in T&D[a] Losses and congestion cost	Deferred Generation Capacity	Deferred T & D[a] Capacity	System Reliability Benefits
Reduction in peak Power requirements	√	√	√	√	√
Provision of Ancillary Services - Operating reserve - Regulation - Blackstart - Reactive power	√	√	√	√	√
Emergency power Supply	√	√			√

[a]T&D = transmission and distribution6.3 DG Barriers.
Redrawn from USDE, 2007a. The Potential Benefits of Distributed Generation and Rate-Related Issues that May Impede its Expansion. A Study Pursuant to Section 1817 of the Energy Policy Act of 2005. http://www.oe.energy.gov/DocumentsandMedia/1817_Study_Sep_07.pdf, 26.05.2010; USDE, 2007b. The Potential Benefits of Distributed Generation and Rate-Related Issues that May Impede Their Expansion. http://www.ferc.gov/legal/fed-sta/exp-study.pdf, 17/04/2013.

Table 6.2   The status of renewable energy technologies during 2007.

Technology	Capacity factor (%)	Turnkey investment costs (US$/Kw)	Current energy cost of new systems	Potential future energy cost
Biomass energy	25–80	900–3,000	5–15 0/kWh	
Electricity	25–80	250–750	1 5 0/kWh	4–10 0/kWh
Heat			8–25 $/GJ	1–5 0/kWh
Ethanol				6–10 $/GJ
Wind electricity	20–30	1,100–1,700	5–130/kWh	3–100/kWh
Solar photovoltaic electricity	8–20	5,000–10,000	25–125 0/kWh	5 or 6–25 0/kWh
Solar thermal electricity	20–35	3,000–4,000	12–18 0/kWh	4–10 0/kWh
Low-temperature solar heat	8–20	500–1,700	3–20 0/kWh	2 or 3–10 0/kWh
Hydroelectricity	35–60	1,000–3,500	2–8 0/kWh	2–8 0/kWh
Large	20–70	1,200–3,000	4–10 0/kWh	3–10 0/kWh
Small				
Geothermal energy	45–90	800–3,000	2–10 0/kWh	1 or 2–8 0/kWh
Electricity	20–70	200–2,000	0.5–5 0/kWh	0.5–5 0/kWh
Heat				

Redrawn from Shrestha, M.R., 2007. Financial Analysis of Renewable Energy Projects. E-Learning Course Bio-Energy for Achieving MDGs, Lecture 7, School of Environment, Resources and Development, Asian Institute of Technology, Thailand. http://www.ofgem.gov.uk/Sustainability/Environment/Policy/SmallrGens/DistEng/Pages/DistEng.aspx 18/04/2013.

3. Short and long term benefits
4. Type of energy source required
5. Type of DG technology needed
6. Health issues

Table 6.3 The costs for distributed generation renewable energy technologies (Redrawn from NREL, 2016).

Technology type	Mean installed cost ($/kW)	Installed Cost Std. Dev. (+/- $/kW)	Fixed O&M ($/kW-yr)	Fixed O&M Std. Dev. (+/- $/kW-yr)	Variable O&M ($/kWh)	Variable O&M (+/- $/kWh)	Lifetime (yr)	Lifetime Std. Dev. (yr)	Fuel and/or water cost ($/kWh)	Fuel and/or Water Std. Dev. ($/kWh)
PV < 10 kW	$3897	$889	$21	$20	n/a	n/a	33	11	n/a	n/a
PV 10 −100 kW	$3463	$947	$19	$18	n/a	n/a	33	11	n/a	n/a
PV 100 −1000 kW	$2493	$774	$19	$15	n/a	n/a	33	11	n/a	n/a
PV 1−10 MW	$2025	$694	$16	$9	n/a	n/a	33	9	n/a	n/a
Wind <10 kW	$7645	$2431	$40	$34	n/a	n/a	14	9	n/a	n/a
Wind 10 −100 kW	$6118	$2101	$35	$12	n/a	n/a	19	5	n/a	n/a
Wind 100 −1000 kW	$3751	$1376	$31	$10	n/a	n/a	16	0	n/a	n/a
Wind 1 −10 MW	$2346	$770	$33	$16	n/a	n/a	20	7	n/a	n/a
Biomass combustion combined heat & power[a]	$5792	$2762	$98	$29	$0.04	$0.02	28	8	$0.04	$0.02

[a]Unit cost is per kilowatt of the electrical generator, not the boiler heat capacity.

7. Security issues (e.g., vandalism/theft)
8. Ownership
9. Location
10. Possible future connection to the main grid
11. Main purpose for installing the system
12. Mode of operation
13. Rating of the DG system
14. The socioeconomic conditions in relation to the local population
15. Training and skills needed for the system operation and maintenance

6.7 Examples of DG technologies

The rapid expansion and development of DG technologies have provided a wide variety of choices, regardless of the size of the project or the geographical location. Since some of the technologies on the market are still in their early stages of development, the

---

BOX 6.3 DG and electric system reliability

DG can be used by electric system planners and operators to improve reliability in both direct and indirect ways. For example, DG could be used directly to support local voltage levels and avoid an outage that would otherwise have occurred due to excessive voltage sag. DG can improve reliability by increasing the diversity of the power supply options. DG can improve reliability in indirect ways by reducing stress on grid components to the extent that the individual component reliability is enhanced.

USDE, 2007a. The Potential Benefits of Distributed Generation and Rate-Related Issues that May Impede its Expansion. A Study Pursuant to Section 1817 of the Energy Policy Act of 2005. http://www.oe.energy.gov/ DocumentsandMedia/1817_Study_Sep_07.pdf, 26.05.2010; USDE, 2007b. The Potential Benefits of Distributed Generation and Rate-Related Issues that May Impede Their Expansion. http://www.ferc.gov/legal/ fed-sta/exp-study.pdf, 17/04/2013.

BOX 6.4 Electrical island

An electrical island is any stand-alone power system, with its own generation and loads, operating in balance.

Sandia National Laboratories, 2013. Suggested Guidelines for Assessment of DG Unintentional Islanding Risk. https://energy.sandia.gov/wp-content/gallery/uploads/SAND2012-1365-v2.pdf. Accessed 14/02/2020.

---

quality and specific requirements are still far from certain. This is why data feedback from current and previous projects in this field is vital. Data feedback can help in further developing these devices in order to meet specific requirements on a long- and short-term basis. These developments can focus on those presently considered as a "complex system(s)," such as a hybrid power system. Most of these systems, usually constructed in a rural area, with limited resources, are in constant need of new ideas, economical design approaches, and flexibility about local resources and the protection of the environment (Boxs 6.3 and 6.4).

6.8 Problems

Apart from the problems related to geographical isolation, the following difficulties can arise:

1. Lack of local technical help
2. Lack of back-up system (or insufficient back-up system)
3. The high cost of repairing and/or device parts replacement

4. The imported device standard may differ from the local/national device standard
5. The time factor, i.e., it may take a long time to send or receive new parts or replacing/repairing faulty parts
6. Whether or not to choose national or imported equipment;
7. Price fluctuation (e.g., fuels, devices, labor, engineering and technical support);
8. Needed financial support may not be available for the local community during the life-span of a system
9. Installing a metering system (or prepaid metering system) can be important for the success of the project and will reduce problems with the local customers

6.9 Conventional energies for off-grid use

In the following chapter, examples of DG devices powered via energy sources, other than renewable sources, have been listed and discussed briefly. These devices are Reciprocating Engines (RE), Micro Turbine (MT), Combustion Turbines (CT), and Fuel Cells (FC). Energy storage systems (see Chapters 1 and 12) are an important part of the DG, in particular when using renewable sources of energy as the main source for generating electricity. Some off-grid systems are also needed in certain hybrid designs, for example, Flow Batteries, Reversible Fuel Cells, Lead Acid Batteries, Superconducting Magnetic Energy Storage (SMES), Ultracapacitors, Compressed Air Energy Storage (CAES), Thermal and Flywheels.

6.10 Conclusion

In this chapter, a brief introduction to DG systems has been examined, together with their applications and various barriers associated with them. The main benefit for a DG system is the supply of electricity to remote areas of the countryside where either it is too costly to connect to the main grid, or in certain cases, not possible to do so. Long-term solutions associated with the electricity supply to the local community should be planned and implemented before and during the actual application of DG systems. Listing various challenges with their possible solutions is one of the approaches recommended during the initial stage of the project planning. Fossil fuels, as part of any hybrid system, should be the last option if it is not possible to use sustainable and environmentally friendly energy resources. Off-grid hybrid systems are one of the best solutions for a more reliable electricity supply in the countryside where certain energy resources are not available or not possible to obtain on a regular basis. DG hardware and hybrid energy systems will be examined in the following chapter.

References

ARE (Alliance for Rural Electrification), Undated. Hybrid Power Systems Based on Renewable Energies: A Suitable and Cost-Competitive Solution for Rural Electrification. http://www.ruralelec.org/fileadmin/DATA/Documents/06_Publications/Position_papers/ARE-WG_Technological_Solutions_-_Brochure_Hybrid_Systems.pdf, 14.06.2010.

Holland, R., Perera, L., Sanchez, T., Wilkinson, R., Undated. Decentralised Rural Electrification: The Critical Success Factors. http://practicalaction.org/docs/energy/Rural%20Electrification.PDF, 29.06.2017.

NREL, 2016. Distributed Generation Renewable Energy Estimate of Costs.

Ofgem, 2013. Distributed Energy - Current Issues.

Shrestha, M.R., 2007. Financial Analysis of Renewable Energy Projects. E-Learning Course Bio-Energy for Achieving MDGs. Lecture 7. School of Environment, Resources and Development, Asian Institute of Technology, Thailand. http://www.ofgem.gov.uk/Sustainability/Environment/Policy/SmallrGens/DistEng/Pages/DistEng.aspx 18/04/2013.

USDE, 2007a. The Potential Benefits of Distributed Generation and Rate-Related Issues that May Impede its Expansion. A Study Pursuant to Section 1817 of the Energy Policy Act of 2005. http://www.oe.energy.gov/DocumentsandMedia/1817_Study_Sep_07.pdf, 26.05.2010.

USDE, 2007b. The Potential Benefits of Distributed Generation and Rate-Related Issues that May Impede Their Expansion. http://www.ferc.gov/legal/fed-sta/exp-study.pdf, 17/04/2013.

Sandia National Laboratories, 2013. Suggested Guidelines for Assessment of DG Unintentional Islanding Risk. https://energy.sandia.gov/wp-content/gallery/uploads/SAND2012-1365-v2.pdf. Accessed 14/02/2020.

Sustainability Victoria, 2011. Distribution Generation Experiences Analysis — Survey Report 2010. http://www.sustainability.vic.gov.au/resources/documents/distributed_generation_experience_analysis.pdf 19/04/2013.

7

Distributed generation systems (DGS)

System hardware

This chapter examines DGS hardware technologies for the purpose of generating electricity, and in some cases, for the generation of heat "cogeneration system" as well. As a continuation of our discussion from the previous chapter, Distributed Generation (DG), or "Generators," is the name used to describe the above technologies. Most of the DGs are constructed close to the location of the energy end users. In general, DGs are used as an emergency power supply, such as in hospitals and commercial buildings. However, within the last 4 decades, DG systems became an increasingly reliable source for generating electricity in remote areas across the globe, but particularly within the countryside in the developing countries.

Renewable energy sources are an essential part of many of the DG systems.

Concerning output, the size of DG systems power can range from 5 kW to 50+ MW, regardless of the energy source.

Benefits of DG systems:

1. Reliability
2. Security
3. Lower impact on the environment than traditional power system (e.g., main grid)
4. Lower costs
5. When using renewable energy sources, DG systems applications may help to reduce greenhouse emissions

This chapter outlines a number of DG systems presently available on the market.

7.1 Reciprocating engines (RE)

These are the most widely used within the field of DG. Generally speaking, there are two types of engines commonly used for the production of electricity:

a. Spark-ignition (Otto cycle)
b. Compression-ignited (Diesel cycle)

For the purpose of generating electricity, the Diesel cycle engine is much more widely used in rural areas. The size of the device can range from less than 1 kW to more than

Rural Electrification. https://doi.org/10.1016/B978-0-12-822403-8.00007-2

50 MW with overall electric efficiencies somewhere between 25% and 50%; however, a smaller unit may not reach the 50% efficiency mentioned above. Benefits may include:

a. Low-cost solutions
b. High/reasonable efficiencies
c. Availability
d. Quick start — switch off option
e. Low operational skill required

On the negative side, these types of engines require regular maintenance, which in turn increases the overall cost. Emissions are also another negative side, even though the emission can be reduced by simply switching to natural gas instead of using diesel fuel. Models available, such as diesel and gas engines (e.g., 5–2000 kW), Compact modular generator (e.g., 4040 kW), Cogeneration units (e.g., electrical output from 70 kW to 2700 kW) (Fig. 7.1).

7.2 Microturbine (MT)

In principle, MT design is the same as any other type of gas turbine. However, the efficiency of MT is much higher than a gas turbine. In fact, some MTs can reuse the exhaust heat to preheat air used in the combustion chamber. The output power can range from 30 kW to or more than 500 kW. Having said that, in spite of a number of benefits from the usage of MT, the costs are higher than the usage cost of reciprocating engines.

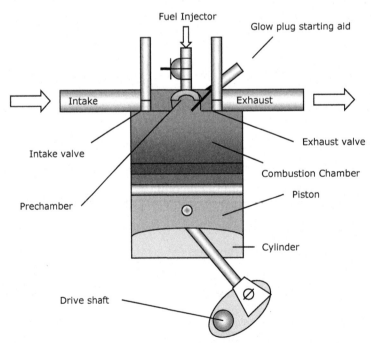

FIGURE 7.1 Schematic diagram of the diesel internal combustion engine.

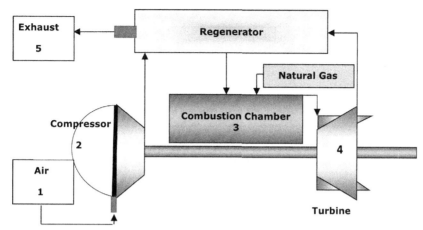

FIGURE 7.2 Schematic diagram of the micro-gas turbine using natural gas.

There are different models of MT designed to use different types of fuels. These fuels can range from natural gas to propane. There are also other types of models that can use biofuels, such as gases produced from landfills (e.g., Methane), and fuels produced from animal waste and/or sewage processing plants. In general, microturbine is a turbine engine with an added generator and electric power in a compact scaled-down size (Fig. 7.2).

The microturbine has a number of advantages compared to other types of DG technologies. MT applications: Cogeneration, peak shaving, base-load power, remote power, premium power, and shaft drive.

There are a number of issues which can delay the expansion of MT: Cost of individual generators in relation to average rural cost, i.e., affordability; additional cost such as utility tariff; financing; grid connectivity; market acceptance; future electric utility regulations and policies (Tables 7.1−7.4).

7.3 Combustion turbine (CT)

CT applications for DG usually range from 1 MW to around 15 MW in their output. However, other types of CT can have an output of around 100 MW. CT is sought after for

Table 7.1 Advantages and disadvantages in using diesel or petrol in a combustion engine.

Diesel cycle engine		Spark ignition engine	
Positive aspects	Negative aspects	Positive aspects	Negative aspects
Can be used for different types of applications	Capital cost is high	Capability of higher running speed	Higher fuel costs with lower efficiency for equivalent power
Fuel price is low	High cost of maintenance	Low maintenance cost	Not suitable for long term usage
High efficiency	Can be large and heavier for a given power	Less heavy, compact, and portable	Range of availability is mostly related to small size
Running speeds are low	Health and environmental risk from pollution	Capital cost is low	Health and environmental risk from pollution

Redrawn and edited from Practical Action (2010).

Table 7.2 Reciprocating engine types by speed available MW ratings.

Speed classification	Engine speed, rpm	Stoic/Rich Burn, spark ignition	Lean burn, spark ignition	Duel fuel	Diesel
High speed	1000–3600	0.01–1.5 MW	0.15–3.0 MW	1.0–3.5 MW	0.01–3.5 MW
Medium speed	275–1000	None	1.0–6.0 MW	1.0–25 MW	0.5–35 MW
Low speed	58–275	None	None	2.0–65 MW	2–65 MW

Table 7.3 List of technologies and nominal system capacity. (Redrawn from the source: US Energy Information Administration, 2016)

Technology	Nominal system capacity
Residential	
Residential — Small solar photovoltaic (<10 kW[a])	5 kW
Residential — Wind	10 kW
Residential — Fuel cell	10 kW
Commercial	
Commercial — Small solar photovoltaic (<100 kW)	40 kW
Commercial — Large solar photovoltaic (100–1000 kW)	500 kW
Commercial — Wind	100 kW
Commercial — Fuel cell	200 kW
Commercial — Natural gas engine	300 kW
Commercial — Oil-fired engine	300 kW
Commercial — Natural gas turbine	1000 kW
Commercial — Natural gas microturbine	250 kW
Industrial	
Industrial — Reciprocating engine	1000 kW
Industrial — Reciprocating engine	3000 kW
Industrial — Natural gas turbine	5000 kW
Industrial — Natural gas turbine	10,000 kW
Industrial — Natural gas turbine	25,000 kW
Industrial — Natural gas turbine	40,000 kW
Industrial — Combined cycle[b]	100,000 kW

[a]Kilowatt-alternating current (kW).
[b]Combined Cycle configuration is (2) 40 MW (MW) natural gas combustion turbines and (1) 20 MW steam turbine.
From Source: Review of Distributed Generation and Combined Heat and Power Technology, 2016. U.S. EnergyInformation Administration. https://www.eia.gov/analysis/studies/buildings/distrigen/pdf/dg_chp.pdf. (Accessed 21 February 2020).

its low emission and relatively low cost. CT also has a high heat recovery and low maintenance requirements; however, on the negative side, CT has low electrical efficiency (ESC, 2010).

Examples of available models such as 300–5000 kW for generators, CHP and Auxiliary Power, 150 MW, Aero-derivative gas turbines, and turbine diesel engines, e.g., from 2 to 124 MW (Fig. 7.3).

Table 7.4 CT commercial availability, performance, and efficiency.

Commercial availability	Size range (kW)	Electrical efficiency LHV (%)	Rural application
Current	500 kW-multi MW	15–40	Yes

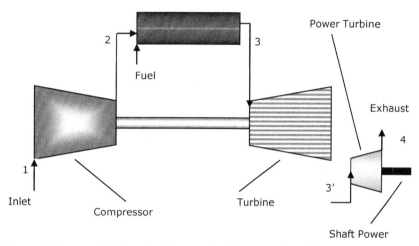

FIGURE 7.3 Schematic diagram of CT showing the operational steps. *Redrawn from ESC (Energy Solution Centre — Distributed Generation Consortium, 2010. Cogeneration. Prime Mover Overview. http://www.energysolutionscenter. org/distgen/Tutorial/TutorialFrameSet.htm. (Accessed 25 April 2010).*

7.3.1 CT summary

1. The principle of CT is that it works in a similar way to a jet engine
2. In cogeneration mode CT is a highly efficient device; efficiency in certain models may range from 60% to 90%
3. CT is commonly used in cogeneration plants
4. Versatile
5. Second lower speed turbine powered by steam in a combined cycle configuration
6. Capacity resource
7. Low fuel cost

7.4 Fuel cell (FC)

Unlike the electricity produced from combustion, the latest fuel cells have higher efficiency than the previous energy devices discussed earlier. FC provides electricity via a chemical process with no waste, no moving parts, and no noise, compared to other devices used for the same purpose. There are different types of FC presently available such as solid oxide (SOFC), phosphoric acid (PAFC), proton exchange membrane

(PEMFC), molten carbonate (MCFC), direct methanol (DMFC), and alkaline (AFC). Apart from the 200 kW phosphoric acid, and up to recently, fuel cells were difficult to obtain commercially, mainly because of the high cost compared with the price of other traditional technologies; however, prices are going down and will continue to do so for the foreseeable future.

Electricity is generated within an FC by splitting the hydrogen atom into a proton and electron.

The protons pass through the electrolyte to the cathode, while the electrons, not able to move to the cathode, will travel around the FC, where a DC current is produced via a second circuit. The voltage generated from FC depends on the number of layers, i.e., the larger the number of layers, the higher will be the voltage output. At the same time, the larger the surface of the electrodes, the greater will be the electrical current obtained (Fig. 7.4).

7.4.1   FC summary

1. Electricity is produced from an external supply
2. Hydrogen and air are the fuel
3. Fossil fuel is needed to provide the hydrogen (e.g., natural gas)
4. Hydrogen and oxygen reaction produces zero carbon emission; however, fossil fuel is already part of the total process, i.e., used during the manufacturing of FC
5. The output current is DC
6. Prices are coming down; however, presently they are still more expensive than conventional energy devices
7. Can be used as a capacity resource
8. The input energy can be more than the output energy

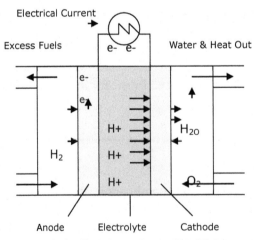

FIGURE 7.4 Schematic diagram of the fuel cell. *Redrawn from USDE (US Department of Energy), 2013. Energy Basic — Renewable Energy. http://www.eere.energy.gov/basics/renewable_energy/ocean_thermal_energy_conv. html 10/05/2013.*

7.5 Renewable energies for off-grid applications

It is very well known that DG powered via renewable sources of energy, such as the small-scale stand–alone off-grid devices, some of them are still in their early stages of development and usage. Sources of renewable energy, such as those obtained from sunlight/heat, wind, and biomass materials, are good examples for harvesting energy using DG devices. An effort to help in providing unlimited sources of energy, supporting the local economy, as well as protecting the environment. Developing sources of renewable energy in rural areas can be difficult to pinpoint one particular type of technology over another, especially in regard to cost and continuity. This is mainly due to individual local conditions, availability, and access to renewable sources of energy. These sources differ considerably in their applications, i.e., hours of services and amount of power required.

As mentioned above, the DG devices using renewable sources are still more expensive to purchase than conventional DG devices. However, renewable energy technologies can be less costly than other types of technologies only when the cost of fossil fuels continues to rise. In addition to the above, many of the rural populations are small and, consequently, per capita demand for energy will be low as well, which may mean further development in renewable energy production is not economical for some of the rural areas in developing countries.

Energy from renewable sources can be summarized under the following titles:

1. Biomass Energy
2. Solar Power
3. Wind Power
4. Geothermal Power
5. Waterpower (including osmosis "osmotic power" and ocean thermal power "Ocean Thermal Energy Conversion – OTEC"

7.6 Examples of RE DG applications

The UN Industrial Development Organization (UNIDO) provide excellent examples of the type of projects needed and the purpose of the energy provided for each rural community, in particular in the developing countries. Examples of various types of projects and their objectives have been selected from the UNIDO publication under the title "Rural Energy for Productive Use." The following projects are from India, Maldives, Sri Lanka, Tanzania, Uganda, Nigeria, and Bauchi. (Note the amount of power produced in each case and the functions providing in the community).

India: Mankulam, Kerala

SHP (110 kW)/Solar Hybrid (1 kW) to power Common Facilities with ICT/VSAT Internet connectivity, Mobile Phone Recharging, Grain Grinding, Cold Storage, Satellite TV, Telephone Exchange and micro-grid for rural lighting.

Maldives: Baa. Atoll

Solar (4 kW)/Wind (6 kW) hybrid to power Community Development Center with ICT/Internet Connectivity, streaming class room programmes to remote islands through internet technology, e-health, e-governance. Solar Thermal Energy for water desalination and hydroponics.

Sri Lanka: Northern, Eastern and Tsunami Affected Areas

SHP (25 kW)/Solar/Wind hybrid systems to power ICT for internet connectivity and other income generating activities in rural, post conflict and Tsunami affected areas in Sri Lanka.

Tanzania: Rukwa Region, Sumbawanga

SHP (75 kW)/Solar hybrid system powered ICT with Internet connectivity, Satelite TV, Community Development Center with micro grid for rural lighting.

Uganda: Bundibugyo, Nduguttu River

SHP (250 kW) powered Community Development Center with ICT/network and internet connectivity, satellite TV, e-health, e-governance, and micro grid for income generating activities.

Nigeria: Enugu

SHP − (30 kW) powered Community Development Center for internet connectivity and income generating activities.

Bauchi

SHP (75 kW) for productive applications including ICT and internet connectivity.

7.7 Solar Power (Solar Photovoltaic Power "SPP")

Solid crystalline materials, such as germanium or silicon (semiconductors), are being used to convert the sun's energy into electricity, i.e., electricity is the output from a solar source by utilizing semiconductor-based materials, commonly referred to as "solar cells." Photovoltaic (PV) generates direct current (DC) of 12 or 24 V. However, by wiring together a number of modules, it is possible to increase the overall voltage output. Similarly, by wiring modules in parallel (array), the overall current output can be increased as well.

As it is not possible to generate electricity round the clock from solar energy, batteries are needed as a backup and storage, as well as to smooth out any irregularities connected to the load requirements, thus enabling the system to be more efficient in its use and function. The batteries cost and their maintenance/replacement will add to the overall cost of the system. Solar-Home-System (SHS) is the most common usage in the majority of developing countries, where a home system power can range from 30−50 Wp with varying estimated costs. In developing countries, the consumer market for PV is growing fast, as is the case with most forms of renewable energy systems. Within the last decade, PV manufacturing growth has increased rapidly. The Compound Annual Growth Rate (CAGR) of PV installations was 36.8% during the years 2010−2018 (Photovoltaics Report, 2019) (Fig. 7.5).

7.7.1   Solar power summary

See Chapter 1 − Section 1.1 Brief Main Aspects of Renewable Energy - sub-Section 1.1.5 (Box 7.1).

7.8 Biomass energy

Reportedly, around 15% of the world's primary energy is from plant materials, and developing countries use around 30%−38% from the same source as a fuel. It is not surprising that biomass provides between 60%−90% of many sub-Saharan African, as well as a number of Asian countries (Altawell, 2014).

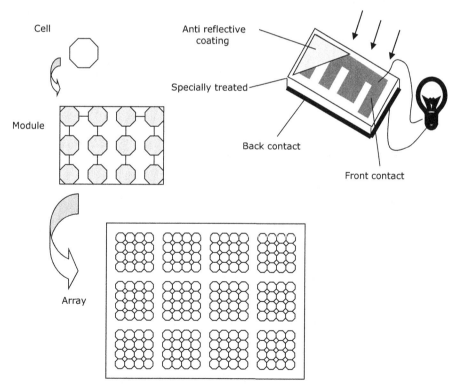

FIGURE 7.5 Schematic diagram of the basic structure and work of solar photovoltaic cells. *Redrawn and edited from Nasa Science (undated). Schematic Diagram of the Basic Structure and Work of Solar Photovoltaic Cells. https://science.nasa.gov/about-us/. (Accessed 14 October 2016).*

BOX 7.1 The over load

"The average load can be easily assessed by logging the current drawn at the main switchboard on an average day. The 'over load' has a different meaning when referred to the D.G. set. Overloads, which appear insignificant and harmless on electricity board supply, may become detrimental to a D.G. set, and hence overload on D.G. set should be carefully analyzed."

From Bureau of Energy Efficiency (undated). DG Set System. Diesel Generating System, pp. 165–178. http://www.beeindia.in/energy_managers_auditors/documents/guide_books/3Ch9.pdf 19/04/2013.

One of the main benefits of using biomass energy crops in rural areas is that it helps agricultural development, i.e., creating commercial agriculture activities and diversification within the agricultural sector when it comes to farming, plus job creation within the local community. Concerning biomass technology, there are two main methods, which cover a wide area of biomass conversion systems, thermochemical conversion, and biochemical conversion. Hardware biomass conversion systems can be stationary or mobile (Fig. 7.6).

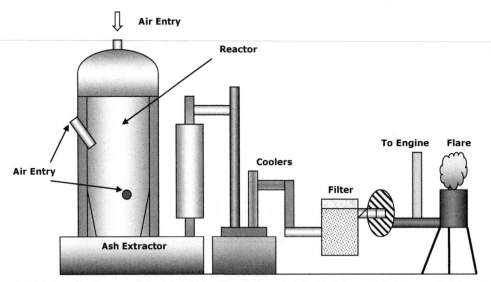

FIGURE 7.6 Schematic diagram of for biomass gasifier. *Redrawn and edited from Global Energy Collaborations, 2010. Technical Details of a 1 mw Biomass Gasifier. Details at biomassgasifier.com. http://www.google.co.uk/imgres?imgurl=http://www.biomassgasifier.com/images/BIOMASS%2520SCHEMATIC%2520DIAGRAM. JPG&imgrefurl=http://www.biomassgasifier.com/TechDetails.htm&h=408&w=528&sz=48&tbnid=8xJFP_URBXBHTM:&tbnh=102&tbnw=132&prev=/images?q%3Ddiagram+of+biomass+for+electricity+production &usg=__qbvLHscnEMZXlGqpgELJ5b2FDxY=&sa=X&ei=3uYhTlfmE5280gTf9an2Dw&ved=0CCAQ9QEwAQ. (Accessed 23 June 2010).*

The mobile hardware systems are usually used in rural areas supplying power for a small number of homes, such as in a village, or powering small to medium size countryside businesses. For rural applications, biomass boiler(s), either as a mobile unit or in a permanent location (or small, modular biomass gasifiers) can be installed using a small system, such as combined heat and power system (CHP) with its main productivity concentrated on electricity generation rather than heat. This kind of conversion means the production of electricity will be around 75%, and 25% will be used either for supplying heat or for cooling systems.

Providing electricity for a medium to large size village, around 1 MW of electricity will be needed. Compared to the conventional method, the saving obtained using the above system can be between 25% and 35%. However, within the last 2 decades, a number of projects using decentralized biomass gasifier systems were implemented in different parts of South-East Asia. The main aim of these biomass projects is to provide electricity for villages, such as in the case of Hosahalli village, Karnataka, in India (Box 7.2).

7.8.1 Biomass energy summary

See Chapter 1 — Section 1.1 Brief Main Aspects of Renewable Energy - sub-Section 1.1.1.

7.9 Wind turbines (WT)

Wind power system requires a control mechanism for the purpose of improving the supply, protecting the system from any possible damage, as well as to optimize the system function and consequently, the output. Generated voltages and system frequency are some of the parameters the above mechanism requires to monitor and control. For rural areas, small wind turbines are used for domestic electricity, village water supply, and irrigation, while micro wind turbines are mainly for domestic electricity supply. Apart from micro-hydro systems, the final cost can be lower than the usage of wind turbines; the wind turbine presently competes with other types of renewable and nonrenewable energy generators.

The most sought after wind turbine system is the one that fulfills the technical requirements of both stand-alone power systems and direct grid feed. Such kind of design

---

BOX 7.2 Biomass conversion technologies

Direct combustion, gasification, pyrolysis, extraction, fermentation.
Anaerobic digestion.

Biomass energy output
Heat and power (CHP), Methanol, Biodiesel, Ethanol, Bio-oil.
Biogas and hydrogen.

makes it adaptable to any new technical development and/or for possible changing of location and/or related to changes within the local environment. In addition to the above, the device should be lightweight, high output in low wind speeds, a wide range of working wind speeds, affordable within the local community in relation to procurement cost and finally, characterized with high efficiency (Figs. 7.7 and 7.8).

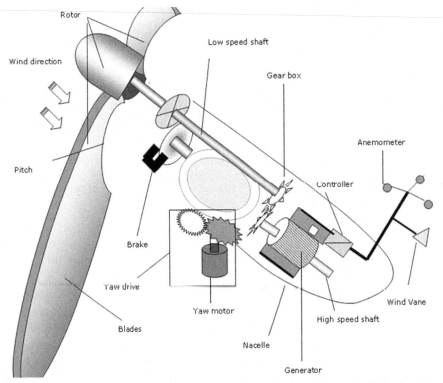

FIGURE 7.7 Schematic diagram of wind turbine parts. *Redrawn and edited from Wind Energy Development Programmatic EIS (undated), Schematic Diagram of Wind Turbine Parts. http://windeis.anl.gov/index.cfm. (Accessed 22 March 2018).*

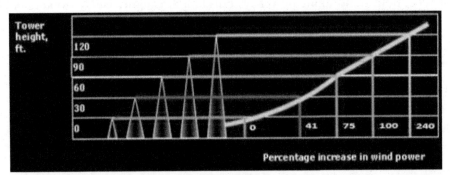

FIGURE 7.8 The wind respectively increases in relation to the height of the tower. *Redrawn and edited from the source: USDI (US Department of the Interior), 2005. Hydroelectric Power — Reclamation Managing Water in the West. http://www.usbr.gov/power/edu/pamphlet.pdf 21/04/2013.*

7.9.1   WT summary

See Chapter 1 — Section 1.1 Brief Main Aspects of Renewable Energy - sub-Section 1.1.6.

7.10 Geothermal

Geothermal energy is distinguished under two headings: High level (or high grade) and low level (or low grade). The first type is mainly related to the super-heated water in the form of steam.

This high-temperature steam is produced under high earth pressure within the deep internal temperature of the earth. Type number two is simply related to the heat originated from the sun within the earth's crust (Fig. 7.9).

The Geothermal Energy Association reported that the number of countries harnessing power from geothermal sources has increased by 120% within the last 2 decades. The report figures indicate that during the year 2000, there were only 21 countries producing power from geothermal sources; however, at present, there are more than 46 countries

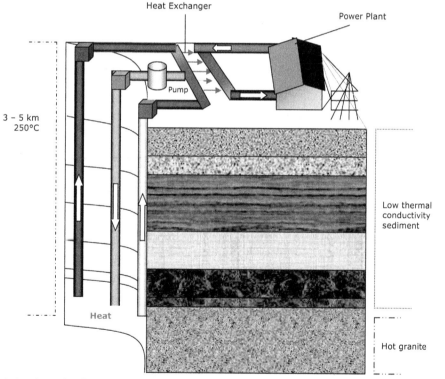

FIGURE 7.9 Geothermal technology supplying electricity. *Redrawn and edited from the Hot Rock Energy program, Australian National University (undated). Geothermal Technology Supplying Electricity the Hot Rock Energy Program. https://www.anu.edu.au/. (Accessed 17 February 2014).*

worldwide. As a consequence of the above, the production of electricity from geothermal sources has grown to 7000 MW in 21 countries during the year 2000 alone (GEO, 2002). Examples: around 23% of the electricity generated in the Philippines is from geothermal sources, and a higher volume of electricity production from geothermal sources can be expected in Indonesia.

According to Fernando Echavarria of the Space and Advanced Technology Office in the State Department's Bureau of Oceans and Environmental and Scientific Affairs (as reported by the *America.gov*. Website 'Amercia.gov, 2008' Actforlibraries, 2017):

"Two regions of the world have the highest potential and yet the least amount of geothermal energy development, one is the East African Rift Valley, which incorporates about 12 countries, and the second is the western margin of South America."

One of the important aspects of geothermal energy, which needs to be considered prior to any planning, is the high cost involved during each step of the project (Fig. 7.10). This kind of cost may not make it possible for producing electricity in an economical way for a small rural community.

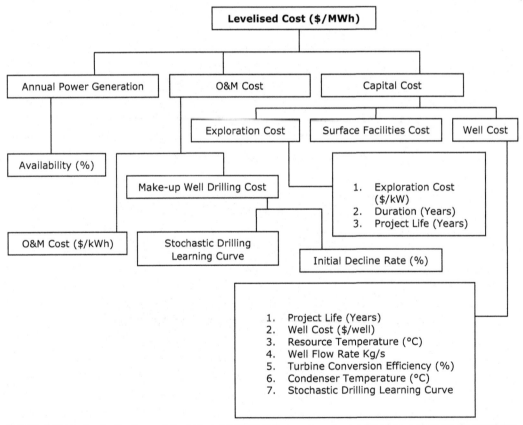

FIGURE 7.10 Stochastic Geothermal Cost Model (SGCM) listing various steps related to a project's financial cost. *Redrawn and edited from Sener et al. (2009).*

Geothermal as a source of energy would be more suitable where the loads balanced with the overall cost will make sense for future cost recovery from a large number of customers benefiting from a geothermal electricity production scheme. Developing countries can benefit greatly from these unlimited sources of energy, particularly where the majority of geothermal sources of energy are located in remote and poor areas of the countryside, such as those in Africa, South America, and different parts of Southeast Asia.

7.10.1 Geothermal summary

See Chapter 1 — Section 1.1 Brief Main Aspects of Renewable Energy - sub-Section 1.1.4.

7.11 Hydropower

Water movement, in general, can be utilized to produce what has been termed as "hydropower." This may range from ocean waves to falling water and underwater currents. Currently, hydropower is the most common way to generate electricity worldwide, in the form of renewable energy. The hydropower system is used to produce electricity by spinning the turbine alternator (generator) or simply for mechanical purposes. The power generated from this source is considered to be the highest on the list in relation to electricity generated from a renewable source. According to the European Small Hydropower Association, 17% of the global supply of electricity is sourced from hydropower, i.e., from already installed generators of 730 GW, and at the same time, 100 GW generators are presently being constructed, worldwide. The objective of the International Center for Small Hydro Power (SHP) (ICSHP), in cooperation with UNIDO, is to provide a platform where hydro projects in rural areas can speed up the development of small hydro power and consequently provide benefit to the local community. For example, the promotion of sustainable development of water resources for the purpose of rural electrification will provide models for other undeveloped hydropower locations in developing their own water energy resources. The training and capacity building in SHP, plus, as mentioned in their final objective (UNIDO, undated), the technology transfer in SHP all add up to the momentum in developing and using hydropower rather than the usage of fossil fuels for the purpose of generating electricity. A number of developing countries obtain a considerable percentage of their electricity via a hydropower source. It has been reported that within the last decade, electricity from hydropower in Kenya is around 55%, Nepal 90% and Peru 48%. As early as 2000, "Hydropower and the World's Energy Future" has reported that the technical feasibility of generating electricity from hydropower in the developing countries, such as in Africa, is approximately 1750 TWh/year and in Asia is around 6800 TWh/year (Hydropower and the World's Energy Future, 2000; Fig. 7.11; Box 7.3).

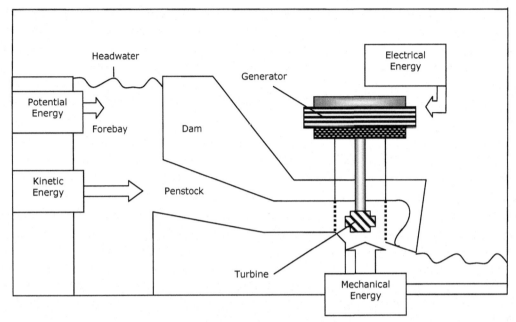

FIGURE 7.11 Example of small hydropower. *Redrawn and edited from USDI (US Department of the Interior), 2005. Hydroelectric Power — Reclamation Managing Water in the West. http://www.usbr.gov/power/edu/pamphlet.pdf 21/04/2013.*

BOX 7.3 SHP main components

Running water/Water conveyance.
Turbine/Waterwheel Generator and Regulator.
Miscellaneous (Wiring, fuses, etc.)

7.11.1   Micro hydropower system (MHPS)

The most suitable hydropower technology for many rural areas is a micro hydropower system (MHPS). A number of projects completed and others underway by the United Nations Industrial Development Organization (UNIDO) for rural micro-hydropower development across many developing countries using basic devices, and cooperation with the local community, has produced excellent results. One of the past successful development is the UNIDO project in Kinko village, Tanzania.

By using natural water supply, such as rivers or streams, MHPS can be easily constructed at a low cost. The work involves a small check-dam, plus the construction of intake wier (to raise the level of water), 9–15 kW generators, mini-grid (for low voltage distribution) forebay as a water reserve/storage (if needed) and water pump as a microturbine (reversed) and batteries, if needed (Fig. 7.12).

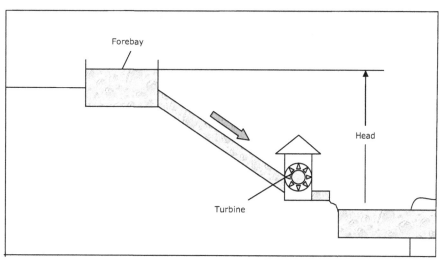

FIGURE 7.12 Example of micro hydropower. *Redrawn and edited from USDE (2010).*

7.11.2 MHPS summary

1. With a small amount of flowing water, MHPS can provide a constant supply of electricity at a very low cost of maintenance (reliability and Efficiency)
2. From small to a large village/community, electricity can be supplied from a small hydropower up to a distance of more than one mile
3. In many cases, there will be no need for a reservoir
4. If the connection has been made to the main grid, then the access of electricity can be sold to the main power station, providing additional income to the community
5. Environmental impact is low
6. The initial cost of purchasing of MHPS parts is low, especially if this cost is compared to the value of energy provided on a long-term basis
7. Suitable locations need to be examined carefully; occasionally it can be a problem or a difficulty in locating the right place
8. In places where amount of water depends on the season, summer time may have an impact on the amount of power produced
9. The amount of water flow may limit future power expansion
10. It can have a good correlation with demand (concerning the amount of water supplied during each season

See also Chapter 1 — Section 1.1 Brief Main Aspects of Renewable Energy - sub-Section 1.1.2.

7.12 Ocean thermal power (Ocean Thermal Energy Conversion - OTEC)

The principle of how OTEC can be used to generate electricity is in the difference of temperature (thermal gradient) between the ocean surface and the deep water. The energy obtained from the difference between these two temperatures can be used to drive a turbine. The thermal ocean gradient should be a minimum of 20°C. There are three different technologies for OTEC system:

1. Closed cycle
2. Open Cycle
3. Hybrid

The "Closed Cycle" system uses the warm surface seawater to vaporize a low boiling fluid (e.g., ammonia), and in turn, the expanding vapors will rotate the turbine. The "Open Cycle" uses low pressure, i.e., using the low-pressure container on warm seawater, the vaporized water will turn the turbine. The hybrid system uses both the "Closed Cycle" and the "Open Cycle" methods, i.e., warm water evaporated under low pressure, which vaporizes a low boiling fluid, which turns the turbine (Fig. 7.13).

A number of developing countries (Tables 7.5 and 7.6) have the natural resources where OTEC can be constructed as a major source of electricity and for desalinated water, i.e., fresh and cold water, which can be used domestically, for agricultural and for industrial usage.

7.12.1   OTEC summary

See Chapter 1 — Section 1.1 Brief Main Aspects of Renewable Energy - sub-Section 1.1.3 (Box 7.4).

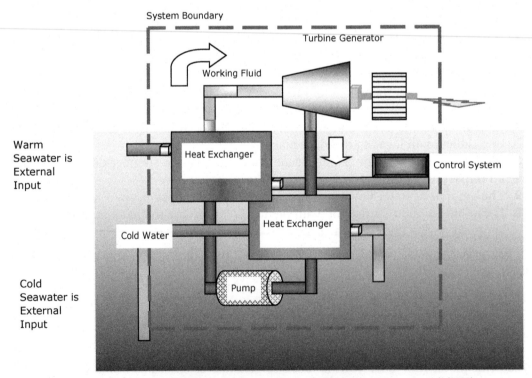

FIGURE 7.13 An outline of OTEC operation. *Redrawn and edited from Cannon et al. (2009).*

Table 7.5 FC performance, efficiency.

Size range (kW)	Electrical efficiency %	Rural application
Several watts to 3000+ kW	35–50	Yes

Table 7.6 Developing countries with ocean-thermal resources (Latin America, the Caribbean, Indian, Pacific ocean, and Africa). Calculated at 25 km or less from shore.

Country or area	Temperature difference (°C) of between 0 and 1000 m	Distance from resource to shore (km)
Latin America and the Caribbean		
Barbados	22	1–10
Cuba	22–24	1
Dominican Republic	21–24	1
Grenada	27	1–10
Haiti	21–24	1
Jamaica	22	1–10
Saint Lucia	22	1–10
Saint Vincent	22	1–10
Trinidad and Tobago	22–24	10
Indian and Pacific Oceans		
Comoros	20–25	1–10
Cook Islands	21–22	1–10
Fiji	22–23	1–10
Kiribati	23–24	1–10
Maldives	22	1–10
Mauritius	20–21	1–10
Samoa	22–23	1–10
Seychelles	21–22	1
Solomon Islands	23–24	1–10
Vanuatu	22–23	1–10
Africa		
Benin	22–24	25
Gabon	20–22	15
Ghana	22–24	25
Kenya	20–21	25
Mozambique	18–21	25
São Tomé and Príncipe	22	1–10
Somalia	18–20	25
Tanzania	20–22	25

Redrawn from NREL, 2010. Developing Countries with Ocean-Thermal Resources (Latin America, the Caribbean, Indian, Pacific Ocean and Africa). Calculated at 25 km or less from shore. https://www.nrel.gov/. (Accessed 14 April 2015).

BOX 7.4 First OTEC

In 1881, Jacques Arsene d'Arsonval, a French physicist, proposed tapping the thermal energy of the ocean. But it was d'Arsonval's student, Georges Claude, who in 1930 built the first OTEC plant in Cuba. The system produced 22 kW of electricity with a low-pressure turbine.

From USDE (US Department of Energy), 2013. Energy Basic – Renewable Energy. http://www.eere.energy.gov/ basics/renewable_energy/ocean_thermal_energy_conv.html 10/05/2013.

7.13 DG using the hybrid system

In order to overcome certain limitations of one or more than one different power system, the solution is simply to integrate these systems in order to obtain the results sought after, which otherwise would not be possible to obtain via the use of one single system on its own. The outcome of power systems integration is one complete system known in the literature as a "hybrid system." The integration may involve electricity generators, storage devices, or fuels (two or more for the same system). The hybrid system is one of the best approaches to solving energy issues in general and electricity generation in particular. This is more so where there is no possibility to connect to the main grid, or there are no facilities to rely on one system or one source of energy. Important factors by which a hybrid system is characterized can be in the form of reliability, flexibility, efficiency, and less emission.

7.14 Hybrid systems

For building a hybrid system, as in the case of any type of DG energy system, information and data will be required. Hybrid systems may combine a number of renewable and nonrenewable sources, thus, for example, the load that the system is expected to meet with higher and lower limitation of energy supply, solar radiation within the chosen location, average wind speed with seasonal and weather pattern and influence, local availability of biomass materials and cost of transport and storage, hydropower in relation to the location, season and the strength of the water flow, fossil fuels and hardware devices (e.g., generators, batteries, boilers, and other similar related devices and parts) with their availability and cost.

Other aspects related to the project should be looked at as well, such as engineering and maintenance procedures and their cost. Wind, Photovoltaic, hydro, fossil fuels, and various types of devices are all integrated in one way or another for the purpose of providing a continuous supply of electricity. Problems related to a hybrid system may arise as a result of complication via the following factors:

1. Uncertain supply of renewable power
2. Load demand

3. The nonlinear characteristics of components
4. The hybrid system sizing and the system interdependent of operations

There are various types of configurations for a hybrid power, where a number of energy sources have been employed at the same time (Fig. 7.14). Taking the above factors into considerations during the early stages of the design, an economical hybrid system, with an optimum output, is possible to achieve.

Finally, the reliability and efficiency of hybrid systems are characterized by making them suitable back-up generators to the main grid, especially in locations where there is a possibility of network failure. In addition to the above, hybrid systems can be vital devices where a regular and continuous supply of electricity is a matter of life and death, such as during hospital surgical operations and emergency wards.

7.14.1   Hybrid system summary

See Chapter 1 – Section 1.1 Brief Main Aspects of Renewable Energy - sub-Section 1.1.2.

7.15 Examples of hybrid systems

There are a number of hybrid systems, which can be used for heat and electricity generation. Examples: PV-wind, Wind-diesel, Wind-PV-micro hydro, and Wind-small hydro.

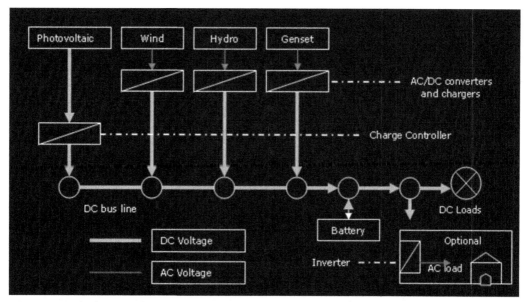

FIGURE 7.14 Hybrid system (DC coupled) combining different RES. *Implementer: STECA (Greece), cited and redrawn from ARE (Alliance for Rural Electrification), (undated). Hybrid Power Systems Based on Renewable Energies: A Suitable and Cost-Competitive Solution for Rural Electrification. http://www.ruralelec.org/fileadmin/ DATA/Documents/06_Publications/Position_papers/ARE-WG_Technological_Solutions_-_Brochure_Hybrid_Systems. pdf. (Accessed 10 October 2015).*

7.15.1   PV-wind (PVW)

There are growing interests in rural areas for small-scale PV-Wind hybrid systems. The phrase "site-dependent" is how wind and solar energy usually described in the literature. However, combining the two sources as one hybrid system will cancel/reduce some of the negative sides of both sources. Although the initial cost is higher in relation to solar and wind systems, when compared to the cost for generating electricity from a diesel engine generator, the maintenance cost can always be lower. Thus a PV-Wind hybrid system reduces the battery bank and diesel requirements, and therefore, from economic perspective, electricity generated in this way can be cheaper than those produced from similar systems (Fig. 7.15). The stand-alone hybrid system is made up of a small PV array and wind turbine. Additional parts needed, such as battery bank, charge controller, and inverter (from DC to AC current).

The estimated cost for generating one kWh by a hybrid system sized for 1.5 and 1.0 kW photovoltaic is less than €0.70, assuming that the life of the hybrid system is 20 years and the batteries bank replacement is every 5 years. The saving cost, particularly on long term usage, i.e., during the life span of the system, is considerable. This kind of saving has been reported by other similar projects in Africa (Box 7.5).

7.15.2   Wind-diesel (WD)

The hybrid WD system is another way of using wind power for the purpose of reducing the dependency on diesel fuel to minimum usage. In order to achieve the above, increasing the number of wind turbines would be necessary. A balance between loads and wind power is important in that saving the amount of diesel being used is part of the saving in term of fuels and maintenance/life cycle of the diesel engine. During the high wind period, not just the diesel generator will stop, but also, any access to wind power

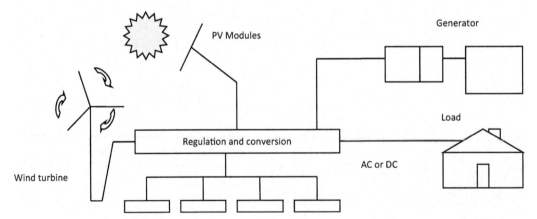

FIGURE 7.15 Example of hybrid power system combining energy sources for the purpose of delivering a continuous supply of electricity. *Redrawn from original source: USDI (US Department of the Interior), 2005. Hydroelectric Power — Reclamation Managing Water in the West. http://www.usbr.gov/power/edu/pamphlet.pdf 21/04/2013.*

BOX 7.5 Solar-PV and wind power components

Basic solar-PV hardware
1. Solar panels
2. Solar panel mounts
3. Solar trackers
4. Inverter
5. Batteries
6. Battery monitor
7. Charge controller
8. Back-up generator
9. AC breaker panel
10. Breaker box
11. DC Breakers
12. Battery and solar panel cables
13. AC and DC wires
14. Miscellaneous (small electrical parts, e.g., fuses, busbars, arrestors, etc.)

Basic wind power hardware
1. Generator
2. Wind controller
3. Inverter
4. Batteries
5. Tower
6. AC and DC wires
7. Back-up generator
8. Miscellaneous (small electrical parts, e.g., fuses, busbars, arrestors, etc.)

---

will be stored for later use. The flexibility, therefore, of the wind-diesel hybrid DG is in the automatic alternating mechanism (real-time) between wind power and diesel generator and stored energy. This kind of design will achieve the optimum in the usage of energy and reduction in total cost. The mechanism, in general, is applied to most hybrid systems where the main controller either specifically written software program for the hybrid system or in certain situations, it can be purchased like any other commercial software (Fig. 7.16).

Wind-Diesel hybrid system economies largely depend on the following factors:

1. The final price (delivered) of diesel fuel
2. Wind power availability
3. Hardware cost
4. Operational and maintenance cost
5. Secondary loads values
6. Demand, revenue, and reliability of the system

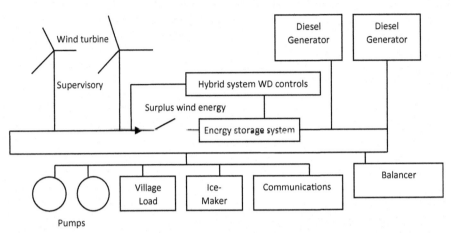

FIGURE 7.16 An example of wind-diesel hybrid system layout. *Redrawn and re-designed from Philippine Hybrid Energy Systems Inc, undated.*

7.15.3 Wind-small hydro

During the year 2010, ESHA and IT published a report indicating that small Hydro Power (SHP) provided electricity close to those produced by other types of renewable energy, which is around 2% of the overall world production, i.e., around 47 GW, where 25 GW of the above source is located in developing countries (ESHA and IT Power, 2010). However, the present growth figures are much higher compared with a decade ago. According to AMECO Research report (Energy and Power) [Global Hydropower Market Analysis 2019 - Projections Report 2026]:

Asia—Pacific region continues to account for the largest share of the global installed hydropower of all regions in the world. China alone accounts for more than 25% of the global hydropower capacity, and added approximately 11.74 GW of new capacity in 2017, including 3.74 GW of pumped storage, taking its total installed capacity to 331 GW, including 26.7 GW of pumped storage. China is expected to witness capacity expansion by 2023, in line with the nation's strategy to minimize reliance on coal. The country has most of unutilized hydropower potential. Therefore, there is a considerable prospect for further hydro development (QYR, 2019).

Developing countries have a vast amount of SHP untapped resources, which can be utilized for countryside electrification. With a low impact on the environment, SHP is a renewable technology, which uses rivers or streams to generate electricity for local usage with an average output power of 10 MW. This amount of power is usually associated with the standard average SHP system, even though this kind of output can be higher in India and China as it can rise between 25 and 50 MW, accordingly.

The cost of installing the SHP system is slightly higher than the cost of purchasing and installing a diesel generator. In the long term, the SHP system is also considered to be cheaper, i.e., there is no fuel needed, neither the associated short-life span, usually diesel engines characterized with. Combining wind power with SHP in one hybrid system is the ideal solution for some of the developing countries (Table 7.7; Fig. 7.17).

Table 7.7 Developing countries with favorable conditions for SHP development.

Region	Short term	Medium term
Asia (excluding India and China)	Nepal, Thailand, Sri Lanka Philippines, Indonesia	Laos, Vietnam
Rest of Asia	India, China	
Latin America	Brazil, Peru	Argentina, Ecuador, Colombia
Africa		Uganda
Caribbean and Pacific		Cuba

Redrawn from the source: Taylor et al. (2006).

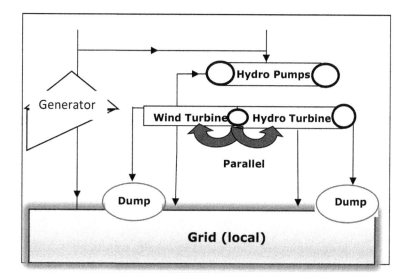

FIGURE 7.17 Schematic diagrams illustrating the basics of the Wind-Hydro Hybrid system.

7.15.4 Wind-PV-micro hydro

The previous three examples of hybrid systems offered a combination of two systems, i.e., either two renewable sources combined or one renewable and one fossil fuel sources combined, as one system. There are three different types of renewable sources being used to form one hybrid system. The idea behind this kind of combination is that not every location is suitable to use it for the purpose of constructing a hybrid system for a continuous supply of electricity all year round. The three sources of energy combined together are where one source (or more than one) cannot be relied on. As has been mentioned previously, this is mainly for factors related to seasonal supply, in relation to the flow of water (hydro) or the amount of solar radiation at a particular time or season, or aspects related to wind power. The three sources of energy, therefore, balance and backup the ways of guaranteeing an uninterrupted supply of energy. The connection and control mechanism is similar in many ways to other hybrid systems, such as those with two sources of energy and/or more than one device/generator.

7.16 Additional technologies

Automated DG functions can reduce operational costs and provide efficiency, accuracy, and reliability. Essential parts to be included in the DG systems, therefore, may include devices, such as distributed control system, fault diagnosis and remote monitoring, metering, supervisory control and data acquisition, protection and control, optimization output and metering. Remote viewing via satellite with continuous data and information feeding/upgrading is one of the best ways to speed up the present and future development in this field (Box 7.6).

7.17 Recommendations

Hybrid systems are the most suitable approach in solving electricity shortages in rural areas, as well as for the reduction of CO_2 emission and protecting the local and global environment. Regardless of the combination of two or more renewable energy systems, the principle is to meet the aim and objective of the DG proposed project for reliable electricity supply at affordable prices. This approach can be implemented anywhere in the world by simply understanding the viability of the energy resources and the practicality of hybrid systems within the local environment.

The recommendations related to DG technologies and sources have been outlined in the following summarized points:

1. Commercial availability of DG devices to the end user in the developing countries, particularly within the rural areas, i.e., manufacturers, suppliers and/or dealers
2. Uniform standard (universal standard) for DG devices and their parts, regardless whether these devices/parts are imported or locally produced
3. Fair competition between the main grid and DG electricity producers should be allowed and encouraged
4. Within a larger community, DG structure should be linked to other similar projects, as well as development schemes
5. Charges for the end user of electricity should be implemented according to the level of employment and the average daily income for each family

---

BOX 7.6 Energy crops

One of the widely reported effects in using energy crops as fuel is the possible stabilization of CO_2 in the atmosphere, which many scientists believe to be one of the main causes of global warming. Apart from the balance needed between the crops used for human and animal consumption and their use as fuel sources, short and long term effects on the environment should be taken seriously, especially when these crops are used for the production of fuels on an industrial scale, globally.

6. DG can, in certain circumstances, minimize demand and reduce peak loads
7. There is a wide choice for DG technologies; however, further development in this field, particularly for devices needed for rural use in the developing countries is still lacking
8. Renewable energy sources (RES) can be one aspect which should be encouraged and supported (via regulations and finance) by the local and national governments, specifically for their applications in rural areas
9. Hybrid systems can be costly at the initial stage of construction; however, the long-term cost saving is one of their important success factors
10. Smart technologies, with or without connection to the main grid, is a vital step in reducing cost and improving the output

7.17.1 Conclusion

Lack of electricity simply means lacking in basic modern essential needs. The countryside of the developing countries is where DG applications can help to speed up the electrification process. In spite of the fact that many of the DG systems are operated by fossil fuels, which is the predominant situation in many developing countries at the present time, the new approach is to support and encourage the usage of renewable sources of energy. International, national, and local government assistance in the form of finance, regulations, and open discussion with the local population, is vital.

Renewable energy sources are plentiful and accessible in many parts of the world, in particular in places where large industrial development has not yet established itself, such as those locations which lack electricity in the countryside.

Systems depending on one source of renewable energy may not be the ideal solution in one location or another, simply because of the unreliability of the source itself and/or the technology is not yet fully developed or matured. DG hybrid systems can provide one of the solutions in speeding up electrification, reducing long-term cost, as well as protecting the environment.

References

Actforlibraries.org, 2017. The Geothermal Energy Potential of East Africas Rift Valley. As Reported by the America.Gov. Fernando Echavarria of the Space and Advanced Technology Office in the State Department's Bureau of Oceans and Environmental and Scientific Affairs website 'Amercia.gov, 2008'. http://www.actforlibraries.org/the-geothermal-energy-potential-of-east-africas-rift-valley/. (Accessed 22 February 2020).

Altawell, N., 2014. The Selection Process of Biomass Nateriqals for the Production of Bio-Fuels and Co-firing. Wiley.

ARE (Alliance for Rural Electrification), (undated). Hybrid Power Systems Based on Renewable Energies: A Suitable and Cost-Competitive Solution for Rural Electrification. http://www.ruralelec.org/fileadmin/DATA/Documents/06_Publications/Position_papers/ARE-WG_Technological_Solutions_-_Brochure_Hybrid_Systems.pdf. (Accessed 10 October 2015).

Australian National University, (undated). Geothermal Technology Supplying Electricity the Hot Rock Energy Program. https://www.anu.edu.au/. (Accessed 17 February 2014).

Bureau of Energy, (undated). Efficiency DG Set System. Diesel Generating System, pp. 165–178. http://www.beeindia.in/energy_managers_auditors/documents/guide_books/3Ch9.pdf 19/04/2013.

Cannon, P., Nandy, S., Nandy, A., 2009. Ocean Thermal Energy Conversion. Slides presentation. Makai Ocean Engineering Inc. http://mason.gmu.edu/~amccull1/files/OTEC_Final_Presentation.ppt#256,1. (Accessed 4 April 2014).

ESC (Energy Solution Centre – Distributed Generation Consortium, 2010. Cogeneration. Prime Mover Overview. http://www.energysolutionscenter.org/distgen/Tutorial/TutorialFrameSet.htm. (Accessed 25 April 2010).

GEO, 2002. Geothermal Energy Facts. (Accessed 25 May 2010). http://geothermal.marin.org/pwrheat.html#Q8.

Global Energy Collaborations, 2010. Technical Details of a 1mw Biomass Gasifier. Details at biomassgasifier.com. http://www.google.co.uk/imgres?imgurl=http://www.biomassgasifier.com/images/BIOMASS%2520SCHEMATIC%2520DIAGRAM.JPG&imgrefurl=http://www.biomassgasifier.com/TechDetails.htm&h=408&w=528&sz=48&tbnid=8xJFP_URBXBHTM:&tbnh=102&tbnw=132&prev=/images?q%3Ddiagram+of+biomass+for+electricity+production&usg=__qbvLHscnEMZXIGqgELJ5b2FDxY=&sa=X&ei=3uYhTIfmE5280gTf9an2Dw&ved=0CCAQ9QEwAQ. (Accessed 23 June 2010).

Hydropower and the World's Energy Future, 2000. The Role of Hydropower in Bringing Clean, Renewable, Energy to the World. http://www.ieahydro.org/reports/Hydrofut.pdf. (Accessed 12 August 2011).

ESHA & IT Power, 2010. Small Hydropower for Developing Countries. European Small Hydropower Association and IT Power. http://www.esha.be/fileadmin/esha_files/documents/publications/publications/Brochure_SHP_for_Developing_Countries.pdf. (Accessed 13 June 2010).

Nasa Science, (undated). Schematic Diagram of the Basic Structure and Work of Solar Photovoltaic Cells. https://science.nasa.gov/about-us/. (Accessed 14 October 2016).

NREL, 2010. Developing Countries with Ocean-Thermal Resources (Latin America, the Caribbean, Indian, Pacific Ocean and Africa). Calculated at 25 km or less from shore. https://www.nrel.gov/. (Accessed 14 April 2015).

Photovoltaics Report, 2019. Fraunhofer Institute for Solar Energy Systems, ISE with Support of PSE GmbH. https://www.ise.fraunhofer.de/content/dam/ise/de/documents/publications/studies/Photovoltaics-Report.pdf. (Accessed 21 February 2020).

Practical Action, 2010. Diesel Engines. The Schumacher Centre for Technologies & Development. http://practicalaction.org/practicalanswers/product_info.php?products_id=34&attrib=1. (Accessed 2 June 2014).

QYR, 2019. Global Hydropower Market Analysis 2019 – Projections Report 2026. Energy and Power. AMECO Research. http://www.amecoresearch.com/market-report/global-hydropower-market-2019-2026-173178. (Accessed 22 February 2020).

Sener, A.C., Dorp, J.R., Keith, J.D., 2009. Perspectives on the Economics of Geothermal PowerICF International. The George Washington University, Engineering Management and Systems Engineering Department, Washington, DC. https://geothermalcommunities.eu/assets/elearning/9.11.GRC2009.pdf. (Accessed 20 July 2020).

Taylor, S., Upadhyay, D., Laguna, M., 2006. Flowing to the east: Small hydro in developing countries. Renew. Energy World 23, 126–131.

UNIDO, (undated). Small Hydro Power: Clean Renewable Water Power. Energy and Cleaner Production Branch. Programme Development & Technical Cooperation.

USDE, 2010. Micro Hydropower. https://www.energy.gov/eere/water/water-power-technologies-office. (Accessed 17 October 2010).

USDE (US Department of Energy), 2013. Energy Basic – Renewable Energy. http://www.eere.energy.gov/basics/renewable_energy/ocean_thermal_energy_conv.html 10/05/2013.

USDI (US Department of the Interior), 2005. Hydroelectric Power – Reclamation Managing Water in the West. http://www.usbr.gov/power/edu/pamphlet.pdf 21/04/2013.

Wind Energy Development Programmatic EIS, (undated). Schematic Diagram of Wind Turbine Parts. http://windeis.anl.gov/index.cfm. (Accessed 22March 2018).

Rural electrification projects

In many parts of the world, rural electrification projects are mostly financed from the public funds, that is to say as part of the governmental schemes usually implemented via organizations such as rural electrification agencies and power utilities. In some cases, the private sectors usually provides the finance needed.

When it comes to the process of financing for rural electrification, varieties of different approaches have to be considered; mostly, they are unrelated to each other in the contest of operation. What this means is that a project providing electricity will examine not just the availability and cost of energy resources but also the social, agricultural, and infrastructural aspects of the rural community. This kind of diversity within the project scope can make it difficult to achieve overall satisfactory results. In addition to the above, financing rural electrification projects can have other negative aspects in that the low population density of the rural community (i.e., low consumption), difficult terrain, low income and the need for subsidies make it unattractive venture for investments.

8.1 Rural projects

The process of obtaining project finance starts with the power utilities approaching an Aid Agency, usually via one of their governmental organizations. The agency will examine the proposal, and the initial approval will lead to the project evaluation. The money offered by the agency can be either in the form of a grant or interest-free loan, or both at the same time. There are, of course, certain terms, conditions, and guidelines, which aid agencies and governmental organizations insist upon before such funding can be made available. Depending on the above conditions, the purchasing of the hardware can be either from the country where the funds originated and/or purchased locally in order to support the local economy where the project is to be implemented. In most cases, the donor country will be the place for selling the plant equipment in order to support their own industries, while at the same time, the aid recipient will benefit from the proposed project.

Other than the UN and the World Bank, for Southeast Asia, a good example where loans and technical help are presently being provided is the Asian Development Bank (ADB). The ADB's task is trying to help two-thirds of the world's poorest people located in the Asia and the Pacific regions by providing grants and/or loans at low-interest rates with long-term technical assistance and financial consultations.

The same thing can be said in regard to Africa, as there are many development banks and aid organizations, which are performing similar tasks in the field of eradicating poverty, as well as for the electrification purposes of the rural communities.

8.2 Electrification projects

To begin the mechanism process for a rural electrification project, apart from obtaining the necessary funds, which may be there at the beginning and/or during the first steps of the project, an initial basic approach to the proposed location would be the starting point. At the same time, working in the background will be the general outline of the proposed work. The initial decisions and steps taken are the most important parts in that they form the basis to build upon as the work develops in the later stages of the project and beyond. Difficulties and unexpected problems, which may arise from time to time, should be dealt with immediately, whenever this is possible, by providing a long-term solution rather than a temporary one.

All solutions provided, before moving to the next stage of the project, should be discussed among the team members, as a way of involving all relevant stakeholders and, for the purpose of highlighting the importance of the project and raise enthusiasm for new ideas and new approaches, while at the same time, trying to cover all angles of the problem(s) connected to the project directly and indirectly.

8.3 Project structure

It is a very well-known fact that a lack of electricity in rural areas can be linked directly and indirectly to a variety of problems, such as deforestation, health care, scarcity of enterprises (unemployment), and so on. In order to provide a platform for financial support to any project, but particularly for a rural electrification project, a sound structure should be in place in the form of practical steps before, or alongside, the launching of an application for financial support.

A basic structural approach, to help achieve the above, can be summarized in the following points:

1. Facts and figures about the proposed location
2. The primary energy source
3. The importance/urgency of electrification for the local community
4. Availability/creation and development of credit schemes
5. Proposed phase plan and period/stages required for present and future development (Figs. 8.1 and 8.2).

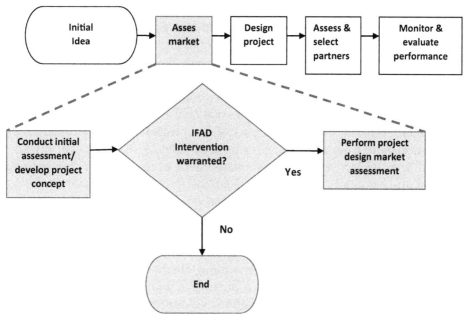

FIGURE 8.1 Market assessment illustrated by using a flow chart. *Redrawn from IFAD, 2010. About IFAD. http://www.ifad.org/governance/index.htm, 8.9.2010.*

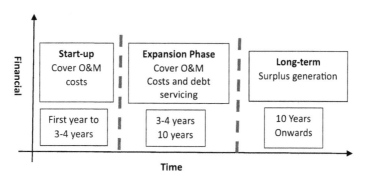

FIGURE 8.2 Predicted time required for financial recovery cost for grid-connected rural electricity supply. *Redrawn from Energy for South Asia, 2004. Subsidizing Rural Electrification in South Asia: An Introductory Guide USAID SARI/Energy Program. Prepared by Contract No. 386-C-00-03-00135-00. Prepared for USAID SARI/Energy Program.*

8.4 Project aspects

According to APEC, 1998, the main headline of the proposed business plan can be summarized as follows:

A. Project identification. This is mainly related to the objectives, management, and experience available

B. **Project Plan.** An outline of the project and related technology, date of implementation, description of the service being offered, agreed location, equipment needed, and infrastructure of the project, human and local resources

C. **Aspects related to marketing.** Overview of the market, customers market segmentation, promotional plan, an initial estimate of the service present and future demand, possible present and future competition in the form of price and regular supply, estimated power capacity, and market share

D. **Financial Aspects.** The overall total cost of the project, estimated fixed assets, estimated revenue from the project in term of sales, revenue stream sales, estimated internal rate of return, risk assessment in the form of the exchange rate, local and national regulations and policy, interest rate, customers power contracts/agreement, estimated other types of sources of finances such as grants - concessional and multilateral, estimated cash flow related to maintenance cost, possible future purchases, service, a default of end-user payment, and other uncounted aspects, which may affect the cash flow

E. **Technical Aspects.** Deciding on the best option whether or not to use local/national or international devices/equipment. Is it possible to assemble components locally? Factors and requirements for the installation. New or second-hand equipment, and if it is the latter, then the operational history should be known. Availability of skilled human resources for monitoring the project. Spare parts and maintenance requirements

F. **Operational Aspects.** Method of charging for the service provided to the customer. Costs related to a transaction. Relevant information for the end user and customers' service guarantees/warranties. Method of payment for the service provided. The standard level of payment. Details of procurement and the way components need to be assembled. Agreed policy on how to deal with defaults on loans and the conditions related to repossession. Training the local community on how to operate the proposed system(s). Local financial personnel who can be used in dealing with customer finance. Method related to the sales, installation, and service

The above criteria form the basic outline, which can be applied to any project, as a means of providing an estimate of the possible financial cost and how it can be adjusted during the process of implementation. Focusing on the developing countries, a few decades ago, the published report "Electricity system performance: options and opportunities for developing countries" (EUR, 1992) pointed out that a number of factors such as social, economic environment, financial and technical aspects should be considered in order to be able to measure any tangible result and performance of the electricity supply. The above factors can still be applied today, regardless of the percentage weighting of importance for each factor in a particular developing country, as well as regardless of how fast or slow these factors changes with time. According to Zomers (2001), the biggest obstacle in some developing countries in providing faster structural development for the purpose of generating electricity is the widespread corruption. Twenty years later, corruption is still a major obstacle to the advancement of these countries. The same author also points out that privatization and corporatization of the power utilities will be needed in order to reduce some of these obstacles, which are constantly hindering rural electrification progress.

8.5 Stages

During the early part of the project's implementation, seminars are an important part of reviewing issues connected to the proposed project model. The seminars should examine the challenges ahead, as well as the opportunities and benefits, which may arise during the work on the project. In addition to analysis of the program management in general and financial aspects in particular, the marketing aspects (assessment) should be taken seriously throughout the period working on the project, and beyond, i.e., including the usage of data to predict future market situations and inclination.

Detailed research of rural electrification policy should be completed before the actual starting of any large project in rural areas (Leno, Undated). This may mean an outlined plan at the start of the project (Fig. 8.2), which then works as a guideline for the financial costing during the early stages. It can work as well as an incentive for potential investors by highlighting the positive aspects of the project. The initial steps taken, therefore, should lead to the eventual preparation of an indicative rural electrification master plan for the proposed project (Boxes 8.1 and 8.2).

Practical measures and stages may include the following (IT Power, 2010):

1. Collecting information on the current status of power supply and generation capacity in close by areas
2. Identification of a rural area as a priority case
3. Further research/study/survey of the location should be carried out to establish the priority needs of the local population

---

BOX 8.1 Project identification

Identified projects can range across the economic and social spectrum from infrastructure to education, to health, to government financial management. The World Bank and the government agree on an initial project concept and its beneficiaries, and the Bank's project team outlines the basic elements in a Project Concept Note.

From The World Bank, 2013. Projects & Operations. Project Cycle. http://web.worldbank.org/WBSITE/ EXTERNAL/PROJECTS/0,,contentMDK:20120731∼menuPK:5068121∼pagePK:41367∼piPK: 51533∼theSitePK:40941,00.html, 10/05/2013.

BOX 8.2 Off-grid project

The conception and implementation of the off-grid project must be consistent with the overall rural electrification plan for the region. The project should not be influenced by such ad-hoc factors as one-time availability of donated renewable-energy equipment or pressure exerted by local politicians, which can be unsustainable.

From The World Bank, 2000. Designing Sustainable Off-Grid Rural Electrification Projects: Principles and Practices. The Energy and Mining Sector Board. http://siteresources.worldbank.org/EXTENERGY2/Resources/ OffgridGuidelines.pdf, 22/04/2013.

---

4. Data obtained from point 3 above should be used to develop a plan for the construction and/or development (e.g., an extension of on-grid and off-grid electrical supply)
5. Assessing sources of energy supply, including the capacity, the engineering and technical feasibility, the operation of the system and maintenance
6. Assessment of possible future demand, including long term strategies for the classification and further progression of rural electrification nearby and/or in different parts of the country, if they are relevant to the present work
7. Locating possible unforeseen needs/problems and strengthening the overall project capacity to make sure that there is a long term sustainable approach for the present and future electrification of the region
8. The promotion of the private sector within the local community and beyond
9. Risk analysis/risk mitigation
10. Call for tenders

To illustrate the final structure of a medium/large team for an electrification project Fig. 8.3 provides an outline of the above.

8.6 Conclusion

This chapter dealt with the basic project structures in order to lead into the other important issues such as those discussed in the following chapters -. finance, method-ology, and energy issues related to the developing countries.

Rural electrification projects are, in essence, no different than any other type of project, whether they are privately funded or financed from the public purse.

Project aspects and project stages have been discussed in order to identify negative and positive approaches. A good practice is to examine the general basic outline of a project, with the provision of possible solutions, before commencing on the main part of the report, i.e., by debating the financial aspects for rural electrifications and providing a robust argument to support the aim of providing a service to the local community.

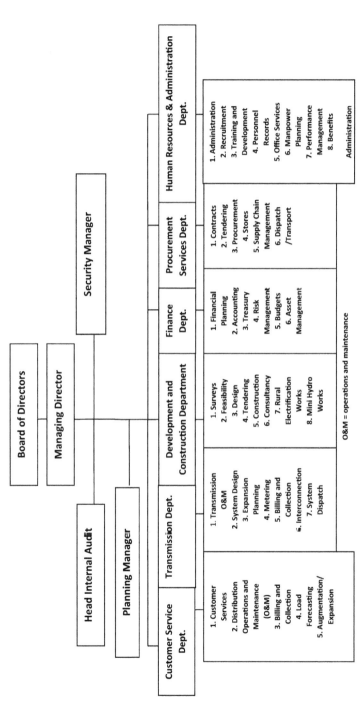

FIGURE 8.3 Example of corporation organization chart in the form of the project workforce and structure. *Redrawn from the Asian Development Bank, 2003. Report and Recommendation of the President to the Board of Directors on a Proposed loan and Technical Assistance Grants to the Kingdom of Bhutan for the Rural Electrification and Network Expansion. BHU 34374. http://www.adb.org/Documents/RRPs/BHU/34374-BHU-RRP.pdf, 31.8.2010.*

However, the structure and the details in every project are somewhat unique, especially when dealing with a large number of changeable factors in remote areas of the countryside.

Mapping these factors with emphasis on the practical applications, cost, energy sources, environmental issues, and the quality of service to be provided is the essence of any successful electrification project.

References

Asian Development Bank, 2003. Report and Recommendation of the President to the Board of Directors on a Proposed loan and Technical Assistance Grants to the Kingdom of Bhutan for the Rural Electrification and Network Expansion. BHU 34374. http://www.adb.org/Documents/RRPs/BHU/34374-BHU-RRP.pdf, 31.8.2010.

Energy for South Asia, 2004. Subsidizing Rural Electrification in South Asia: An Introductory Guide USAID SARI/Energy Program. Prepared by Contract No. 386-C-00-03-00135-00. Prepared for USAID SARI/Energy Program.

EUR, 1992. Electricity System Performance Options and Opportunities for Developing Countries. EUR 14577 EN. Luxembourg.

IFAD, 2010. About IFAD. http://www.ifad.org/governance/index.htm, 8.9.2010.

IT Power, 2010. Rural Electrification. http://www.itpower.co.uk/Services/International+Development/Rural+Electrification, 15.9.2010.

Leno, V., Undated. New Approaches in Rural Electrification Fund Design for Sustainable Development - the New Paradigm. Paper Presented at the Regional Conference on Enabling Environment for Private Participation in Rural Energy Service Delivery and Financing in the SADC Countries.

The World Bank, 2008. Designing Sustainable Off-Grid Rural Electrification Projects: Principles and Practices. The Energy and Mining Sector Board. http://siteresources.worldbank.org/EXTENERGY2/Resources/OffgridGuidelines.pdf, 22/04/2013.

The World Bank, 2013. Projects & Operations. Project Cycle. http://web.worldbank.org/WBSITE/EXTERNAL/PROJECTS/0,,contentMDK:20120731∼menuPK:5068121∼pagePK:41367∼piPK:51533∼theSitePK:40941,00.html, 10/05/2013.

Zomers, A.N., 2001. Rural Electrification. Ph.D. thesis. University of Twente. Twente University Press. http://doc.utwente.nl/38683/1/t0000008.pdf, 13.8.2010, 211.

9 ▪▪▪
▪▪▪
▪▪▪

Finance

Financing electrification projects successfully requires the understanding and implementation of certain factors connected wholly or partly to the make-up and functionality of the project. Among these factors are aspects related to project affordability, reliability, sustainability, and present, as well as future development.

A variety of basic financing aspects will be discussed and dealt with in this chapter, regardless of the original aim(s) and objective(s) of the project, either for the purpose of establishing off-grid electricity supply (in the form of renewable energy system) and/or to extend the supply from the national grid to the town or the village concerned. For the above reasons, a number of financial aspects, as well as financial institutes/organizations with their resources and structures, will also be discussed. The purpose behind the discussion is simply to provide a window to the world of finance, in particular, where rural electrification is concerned.

9.1 Rural finance

It is well known that the flow of finance to rural areas differs in many ways from that of urban cities. This is mainly due to the priority given to a densely populated location by local and national governments over those sparsely populated parts of the countryside. This kind of "priority" "approach may have been the logical theme among many planners, investors, and governmental organizations in the past. However, the development of technologies in various fields and the adoption by many countries of international regulations related to global warming, make the above priority unjustified, particularly for long-term development prospects.

The sources of finance in developing countries for rural electrification do not differ from those obtained for the urban areas. The funds can be sourced from the annual budgetary allocation, from international financial organizations, levies on electricity sales, fuel levy, and levies on specific isolated systems, including systems for private consumption. As it has been mentioned previously, in many parts of the world, rural electrification is mostly financed from the public funds, i.e., part of governmental schemes implemented via organizations such as rural electrification agencies and power utilities.

There are also government-sponsored borrowing schemes carried out either via independent financial institutions or in the form of a new scheme created by the central or local government with the participation of one or more agencies. In certain cases, the private sectors may provide the finance needed, encouraged by various governmental incentives such as grants, lower taxation, and supportive regulations. In this chapter, an outlook has been provided concerning rural electrification sources of finance in general with an emphasis on various parts connected to financing mechanisms such as approaches, banking, process, challenges, and solutions. Examinations of the methods used as a way of tackling some of the problems connected to this field of work will focus on some of the methods used for rural finance, such as microfinance and the cooperative approach.

9.2 The process

When it comes to the process of financing for rural electrification, a variety of different approaches have to be considered; mostly, they are unrelated to each other in the contest of operation. What this means is that a project providing electricity will examine not just the availability and cost of energy resources but also the social, agricultural, and infrastructural aspects of the rural community. All the above approaches have to be considered from different angles before a final decision can be made on the final allocation of the funds. This kind of diversity within the project scope, in relation to the special features of a rural electrification scheme, can make it difficult to achieve satisfactory overall results, in general, especially when trying to involve the private sector. In addition to the above, financing rural electrification projects can have other negative aspects in that the low population density of the rural community - meaning low consumption - difficult terrain, low income, and the need for subsidies make it an unattractive venture for investments, i.e., regardless of whether the finance is sought from the private sector or even from public funds. With regard to India's public finances, the Indian governmental source of finance and mechanism available for financing projects, including off-grid power generations in the form of renewable and nonrenewable energy projects, could follow either:

- A traditional method where the state government provides the funding either from the available development funds or the state government may provide the finance via official governmental borrowing.
- Central Sector Projects will be the starting point in that the office of the government development budget, as well as the office dealing with the official borrowing, will deal with the financing aspects. The official borrowing office deal with multilateral, bilateral, commercial banks, and suppliers' credit. Both methods will end up dealing with the power utilities, where a power project, among other projects, will be dealt with according to priority and cost. Therefore, a financing process may start with the power utilities approaching an aid agency, via one of their governmental organizations. The agency will examine the proposal, and initial approval will lead to the project evaluation. The money offered by the agency can be either in the form of an interest-bearing loan, a grant of free interest loan (or can be both at the same time). Other than the UN and the World Bank, for South Asia, a good example of where loans and technical help are presently being

provided is the Asian Development Bank (ADB). The ADB's task in trying to help two-thirds of the world's poorest people located in Asia and the Pacific region by providing grants and/or loans at low-interest rates should expand their services for further funding when it comes to rural electrification in general and off-grid in particular.

9.3 Financing issues and barriers

Financial institutes, especially large multilateral and bilateral development banks, can be complex to deal with, and typical barriers, can thus be summarized as follows (IEA, 2004):

1. Obtaining finance from some of the development banks can be a complicated and long process
2. Financing institutes may view the majority of the projects as small in size, and at the same time, face high transaction costs for investment preparation and financing
3. Many of the large international finance providers deal only with national governments (or their representatives) and/or with projects where the national government is involved
4. The majority of the above financial institutions have their own set of conditions, regulations, and guidelines, which should be met before the proposed application for finance can be processed fully
5. If the system or technology used in the proposed project is not fully accepted/established (e.g., unfamiliarity with the technology/high risk) by the present market and/or not yet fully proven, then the financing may not be possible
6. Future returns on the loan and high profitability may not be an incentive to the lender if the requested finance is not large enough
7. In certain situations, the competition to obtain a grant or loan from limited funds could make the process for project financing much harder, especially if not enough research and preparations were done before presenting the case for finance to the public and/or private financial institution

In addition to the above, there is another obstacle in accessing finance for rural electrification in that potential customers expect their electricity supply to be similar to those in urban areas, i.e., similar to those obtained from a centralized electricity supply, which, of course, may not be the case. This kind of expectation has been confirmed by the UN Department of Economic and Social Affairs (Division for Sustainable Development) in their report titled "Case Study of a Successful National Energy Program/Strategy: Rural Electrification from Renewable Energy and from Other Sources."

9.4 Solutions

The solution to the barriers mentioned in the previous section can be outlined in the following points (Parliamentary Office of Science and Technology, 2002):

1. A program that can deal with structural adjustment and energy reform
2. Privatization of public energy utilities
3. Removal of subsidies

However, when it comes to the laws and regulations of a particular developing country, there are a number of points, which should be dealt with within the board of management of the project. At the same time, these points should also be raised in relation to project finance and implementation (Leno, Undated):

1. Funds for the rural electrification should be regulated within the legal framework of the country where the proposed electrification project will be implemented
2. More effective and wider responsibilities for the government ministry (or ministries) in charge of energy issues
3. Representations from various governmental, private agencies, as well as from those offices dealing directly and indirectly with the proposed project, should be included within the financing approach
4. In addition to being accountable and transparent in their day-to-day functions, their work itself should be accomplished on a commercial basis rather than institutional/governmental rules
5. For attracting funding, project operational tasks should be fully in the hand of the chief executive (or director) of the project. This kind of procedure will make it easier in raising further funding for the proposed project
6. International laws and regulations should be an essential part of the policy and financial implementation
7. The private sector and local institutions/organizations should be involved, particularly when aiming for a long-term solution
8. Active and deeper financial cooperation between the public and the private sector is necessary
9. Financing for RE projects should be considered on the basis of the size and the requirement related to the project. In some cases, large to medium size projects are usually connected to the main grid, where commercial companies (or independent power producers) operate. Financial risks noticed by the investors for projects such as these are in the form of credit record, which these new projects do not possess; neither they have experience in the field, compared to well established larger power suppliers. Credit guarantees by the government, in such cases, will be needed as part of the solution for attracting further funding

10. Concessions are necessary from the local and national government to private investors and cooperative organizations in order to encourage investment in small projects. These concessions can be in the form of tax relief, for instance, according to the amount of financial investment provided and the actual implementation for each stage of the project
11. Introducing reliable long term RE/Off-grid national and local policy by individual countries will make investments in the above field more attractive and less risky
12. Political stability (in regard to what has been termed as *fragile states*) in developing countries is vital for attracting the finance needed, as well as for the purpose of implementing and developing rural electrification projects successfully in these countries

9.5 Financial structure

Successful rural electrification financing can be rooted within the financial structure of the country concerned. This means that a well-established financial structure, in any country, can be the key for faster and better solutions to a variety of rural and urban electrification projects. Three structures have been looked at briefly below, then checked in relation to RE financial investment in the countryside, regardless of the type of the power system needed, i.e., off-grid or connected to the main grid.

9.5.1 Micro-level

A. Financial services provider and their customers
B. May include households and individuals, regardless of their financial status
C. Commercial banks, agricultural development banks, postal banks, and postal savings banks, financial cooperatives, credit unions, NGOs that provide financial services, agricultural supply agents, and insurance and leasing companies.
D. Indirect influence on meso- and macro-level.

9.5.2 Meso-level

A. Financial-sector infrastructure and support services
B. May include domestic rating agencies, credit information bureaux, audit firms, deposit insurance agencies, training and technical service providers, professional certification institutes, and the networks, associations and apex organizations of financial service providers
C. Reduce transaction costs, improve sector information and market transparency, increase access to refinancing and enhance skills within the meso-level
D. Facilitate activities in the financial sector and not providing retail financial services
E. Direct and indirect influence on micro- and macro-level

9.5.3 Macro-level

A. Legislative and policy framework of the financial system
B. Governments, central banks, and ministries of finance
C. Financial law, supervising financial institutions, and enforce compliance
D. Responsible for the overall economic conditions of the country
E. Indirect impacts on the financial system via governmental macroeconomic policies, such as monetary policy
F. First, direct and indirect influence on meso- and micro-level

Second, to facilitate investment in RE in general and off-grid projects in particular, there should be sufficient supportive factors within the following levels:

9.5.3.1 Micro-level
1. Sufficient national legislations for the purpose of regulating the uses of renewable energy sources
2. Clarity of laws for all aspects related to renewable energy promotion
3. Strong capital market in general and RE finance in particular

9.5.3.2 Meso-level
1. Complete reform of the power sector
2. Where RE could have similar, if not better, competitiveness outlook
3. The above should cover local and national approach, i.e., in order to enable RE to compete with other types of power supply successfully

9.5.3.3 Macro-level
1. Political stability
2. Governmental practical financial and regulatory support
3. Active and growing national income

9.6 Financial schemes

Financial schemes differ from one country to another, as does the percentage of the adult population who are able to have access to financial institutions. In Table 9.1 is an illustration showing less than half of the population in the majority of countries listed in the Table do not have access to financial services. This, of course, could depend on the ability of the potential customers to meet their bills, the availability of the credit schemes such as banks and indigenous organizations, and the availability of soft loans and/or grants (ELC – Electroconsult, Undated).

Financing for off-grid projects can be made via the three different main sources of finance (Table 9.2; IEA, 2004):

1. International concessionary financing
2. National development financing
3. Commercial financing

Table 9.1   An estimated percentage of the adult population accessing financial services. (The World Bank, 2007).

Country	%
Brazil	71.04
China	80.23
Egypt	32.78
India	79.88
Jordan	42.49
Indonesia	48.86
Vietnam	38.80
Pakistan	21.29
Afghanistan	14.89
Iraq	22.67
Mexico	36.93
South Sudan	8.57

Point 1 - covers a variety of international financing multilateral development banks, where loans or grants can be made available for rural electrification projects and other areas of national and local development, mainly for the purpose of reducing poverty. Loans can be provided in the form of "soft" loans, subsidized, or loan offered at a commercial rate. Banks can also provide guarantees where other avenues of financing have been approached.

Point 2 - A budget allocated for national and local funding can be the source of finance for new and established development projects. The majority of development projects, including rural electrification projects, can be financed wholly or partly from the above budget. International funding institutes may contribute to financing as well, in partnership with the local and/or national government.

Point 3 - Sources of finance can be obtained from commercial banks if and when convincing presentations have been demonstrated to show future high-profit returns. These banking institutions with large financial resources can be available at various levels, i.e., locally, nationally, and internationally.

9.7 Financing stand-alone (off-grid) systems

Generally speaking, the off-grid lending program in developing countries can only be successful if there is an established partnership between the energy business and the lending body (e.g., microfinance institution).

There are hesitations (or even outright rejection) from various lending organizations when the issue is raised concerning borrowing money for an off-grid electricity project. This is because the project may not be a familiar thing to the financing body concerned, and consequently, a negative response can be the norm. For this reason, other

Table 9.2 Illustration related to sources of financing tools.

	Market-based loans	Soft loans	Grants	Equity investment	Guarantees
Multilateral development banks	√	√	Some	√	√
Bilateral aid	√	√	Some		
Funds/foundations	√	√	√	Some	
Green investment				√	
National development funds	√	√			√
Commercial loans and investment	√			√	√

Redrawn and edited partly from the source: IEA, 2004. Sources of Financing for PV-Based Rural Electrification in Developing Countries. Photovoltaic Power Systems Programme. Report IEA-PVPS T9-08:2004. http://www.iea-pvps.org/products/download/rep9_08.pdf. 20.12.2010.

supporting incentives should be in place when approaching a lending institute, e.g., guarantee or partial risk guarantee should be offered to the lender in order to reduce possible risk.

Within the focus of Special Projects Manual, according to MME, 2009 (cited IEA1, 2010), those alternative electricity sources, not fossil fuel-based, receives 85% of the electrification costs as grants, while the rest (15%) will be financed by the company providing the electricity.

The total funds released are in accordance with the timeline (Table 9.3). In regards to renewable energy off-grid projects, in spite of the fact that there is little difference between financing a project for rural electrification where fossil fuels are the main source of energy or renewable sources, there are additional difficulties for obtaining finance when renewable energy projects are discussed. This is due to three main reasons associated with policy, finance, and institutional issues. The financial aspects related to

Table 9.3 Example of stand-alone systems timeline funds release.

Quota	Condition	Release of funds (% of the contract)	Accumulated release (%)
Initial release	After signing the concession contract and in compliance with all legal requirements	30	30
2nd release	Subject to financial accountability of the quota anticipated at the contract signing, as well as the confirmation of purchase and receipt of materials and equipment	Up to 60	Up to 90
Final release	After proof of financial performance and physical inspection - to be held after receiving the final commissioning report by the service providers	Up to 10	Up to 100

Source: Cited IEA, 2010. Energy Subsidies: Getting the Prices Right. http://www.iea.org/files/energy_subsidies.pdf. 13.9.2010, from the original source MME, 2009. Programa Nacional de Universalização do Acesso e Uso da Energia Elétrica – Manual de Projetos Especiais. Anexo à Portaria No 60, de 12 de fevereiro de 2009. Eletrobrás/Ministry of Mines and Energy, Brasília.

RE can range from limited access to equity, lack of bankable projects, transaction costs, and availability of local banker lending for RE, as well as the difficulties in obtaining grants from donors/countries when RE projects are discussed. At the same time, there are other aspects influencing financing RE projects, such as institutional issues, i.e., the lack of capacity to develop, implement, and operate RE projects, in addition to the energy links to other sectors.

Policy issues, such as policy bias toward fossil fuels, artificially low end-user prices by utilities, inappropriate or absent regulatory frameworks, the deregulation of the energy sector, and lack of policies for supply to rural areas.

The variety of research conducted on how to implement and develop RE in rural areas in developing countries and the citation of social, technical, economical, and institutional factors (Liming, 2009) indicate that the main major obstacles for the RE are:

1. The lack of finance in backing RE projects, insufficient acceptance of RE
2. Lack of standard and capacity, i.e., not many businesses operate electricity systems with renewable resources
3. Lacking the required infrastructure
4. Majority of entrepreneurs do not have the necessary skills
5. The high initial costs of photovoltaic systems, which can make it beyond the financial capability of the poorer section of the community
6. The unavailability of technical and related information concerning renewable energy sources and the absence of reliable solar radiation data, as well as the scarcity/availability of RE hardware on the market
7. No formal technical standards, in particular, for photovoltaic equipment and components, nor the availability of instructions for installing and maintaining photovoltaic systems
8. Lack of instructions on basic operation and maintenance, and finally, lack of project developers in this field.

However, the following points can accelerate the financing prospect for renewable energy:

1. Financial incentives in the form of grants, tax relief/exemption, subsidies, project priority consideration/allocation, and support
2. Financing and making available RE hardware in the local market
3. Financing training schemes for the local population for various aspects connected to RE, including awareness-raising program
4. Creating, supporting, and monitoring local credit schemes involved directly with RE projects

In general, the question here is how to finance a rural electrification project, in particular RE projects.

According to Kalra et al. (2007), the main funding in India for rural electrification can be sourced from:

A. Corporation specialized in rural electrification projects
B. Plan allocation to the states
C. Direct Funding from the government (e.g., in the form of loans or grants)
D. Commercial banks
E. International finance institutes

In regard to point C above, steps have been introduced by the Indian government for a host of financial incentives. Some of these incentives can be summarized in the following points:

1. During the first year of system installation, there will be 100% accelerated depreciation for tax purposes
2. Lower import tariffs for remaining hardware and parts
3. There is a 5-year tax holiday for power generation projects
4. Where power is generated via the RE system and fed to the grid, the government will provide "remunerative price" under alternate power purchase policy
5. Banking facility plus wheeling of power
6. Subsidies and/or financial incentives for expensive renewable energy devices

9.8 Sources of finance for rural electrification

9.8.1   UN (United Nations)

If comparisons made between the spending budget for some of the developed countries and the UN budget, then the UN budget is considerably smaller, especially considering that the UN is one of the biggest and most important international organizations. Funds spent for a variety of international projects, as well as for the overall work by the UN, have averaged around $30 billion per year up to 2010 (Global Policy forum, 2010). In spite of the constant UN budget deficit, UN funds for rural development contributed to various development projects, including those connected directly and indirectly with rural electrification. In fact, the majority of UN departments, in one way or another, contribute to poverty reduction in the form of financial schemes, mostly allocated to developing countries. These financial schemes include funding for rural electrification projects.

Financing tools and approaches to fund and support RE projects by the UN have been presented in various forms, such as the Sustainable Energy Finance Initiative (SEFI) by the United Nations Environmental Program (UNEP). The aim of SEFI is to encourage investment in the field of RE projects. The support from UNF (United Nations Foundation) for SEFI for support of RE in various fields, particularly where RE finance is concerned, is an important step for pushing forward RE applications for rural areas in the developing countries. The following brief discussion looks at various sections of the UN financing methods, which they are connected directly and/or indirectly with rural electrification, in general, and RE in particular.

9.8.2   UNCDF (United Nations Capital Development Fund)

In order to provide a strategy that can help to provide access to a variety of financial sectors for the poor and low income in the developing countries, the UNCDF believes that financial sectors need to be supported in the form of appropriate policy with legal and regulatory frameworks. In order that this approach can be achieved, the same report recommended the following summarized points:

1. Access should be made available to all kind of financial services at reasonable cost
2. Standards, accountability, and transparency should be part of the tools guiding the financial institutions at every level
3. Sustainability for financial and institutional organizations in order to guarantee long term financial services
4. A variety of financial services providers should be encouraged, whenever this is possible, in order to create competition and consequently cost reduction and options for the customers

The above include funding for renewable energy projects, especially where the lack of electricity supply is the main concern.

9.8.3   Supporting financial services

The UNCDF is in a unique position as it operates as an investor by putting its resources within the macro, meso, and micro schemes. In addition to the above, the UNCDF presents itself as a neutral facilitator for a range of actors. The availability of the UNCDF technical staff at various levels makes it among one of the best institutions in supporting financial services. According to the UNCDF, the organization focuses on the largest constraint, which it believes to be lack of retail capacity. In regard to microfinance, the UNCDF is engaged in active dialogue with those investors who are looking for further methods of investments beyond the traditional banking market. For this reason, within the national structure of a country, UNCDF support at the meso-level mean in itself support to national microfinance associations for their work in providing financial services to the local community. At the same time, the UNCDF provides help to the Central Bank capacity by working with donors and associates in various ways connected to regulatory issues as well, as in the form of training.

Finally, the promotion of the financial sectors can be summarized under the following points:

A. There should be a dedication for the best practices, as has been stipulated in the *"Good Practice Guidelines for Funders of Microfinance"* (CGAP, 2006)
B. Financial outlook when it accepts risk investment in new businesses and postconflict projects monitoring results
C. To successfully accommodate growth by involving other funders

D. Monitoring the wider outlook for microfinance funders in various parts of the world
E. The UNCDF Work with other agencies of the UN, in particular when it comes to technical aspects
F. UNCDF support innovation with the main aim of supporting financial services to the poorer community
G. Training and funding for the UNCDF staff is a priority as part in improving the organization services
H. Learning from past and present work experience, as well as via the active cooperation with its partners and technical staff
I. An approach to involve the private sectors

9.8.4   UNDP (United Nations Development Program)

As a global development network, the main role of UNDP is to help the global community achieve a better life via the sharing of knowledge and experience of all their member countries.

One of the Millennium Development Goals made by the world leaders is to reduce poverty, worldwide, by the year 2015 (Environment and Energy, UNDP, 2010), Presently, at the time of writing this book, we are in the year 2020 and poverty is on the rise in certain countries rather than declining. At the same time, the UNDP mentioned in the above report that it will work on aspects related to *"Environment and Energy," "Crisis Prevention and Recovery," "Democratic Governance"* and areas related to health aspects, such as HIV/AIDS/CORONAVIRUS. In addition to the aspects related to human rights and the empowerment of women across the globe, the financial fieldwork of the UNDP focuses on developing and attracting aid to underdeveloped countries. The HDR (Human Development Report) annual report believes that the main issue is the speeding-up of the development of the poorer nations, in addition to other global issues of development.

9.8.5   Environment and Energy

Energy is central to sustainable development and poverty reduction efforts. It affects all aspects of development – social, economic, and environmental – including livelihoods, access to water, agricultural productivity, health, population levels, education, and gender-related issues (Environment and Energy, UNDP, 2010)

The above is part of the opening statement of the UNDP on the website for the Environmental and Energy program. The UNDP is trying to create an energy policy focusing on rural areas in the form of developing the local capacity and facilitating consultancy. At the same time, the policy is connected to the provision of energy service to the poor, which is part of the main aims and objectives of this section of the UNDP.

The Environment and Energy program of the UNDP focuses on its work on the following points:

A. Sustainable development
B. Water aspects
C. Access to energy (sustainable)
D. Land management (sustainable)
E. Conservation/biodiversity (sustainable)
F. Emission control (National/sectoral)

9.9 The World Bank

As one of the UN specialized agencies, the World Bank was formed by the 184 member countries and made decisions collectively on how the bank should operate. The word "Bank" should not be taken in the meaning of the commercial high street bank, but rather as an important development agency.

In regard to some of the developing countries, many of the world's poorer countries are not able to use the international market to obtain loans for their internal development. This is mainly due to the high-interest rates being charged on these types of loans. The World Bank's role is to extend to these countries financial services by either providing interest-free loans, grants, and/or loans at a low-interest rate.

According to the World Bank, during 2004, $20.1 billion was provided for a variety of projects (around 245 projects). The repayment of loans, usually with the option of 10 years grace period, can be made between 35 and 40 years later, according to the conditions stipulated in the contract. In addition to the finance provided, the World Bank can offer technical assistance in various fields. This kind of approach by the bank is mainly to help reduce poverty by supporting the basic development structure of electricity projects, providing clean water supply, health issues, and protecting the environment. The World Bank usually functions mainly as a service provider for the developing countries in the following banking fields:

1. International Bank for Reconstruction and Development (IBRD)
2. International Development Association (IDA)
3. International Finance Corporation (IFC)
4. Multilateral Investment Guarantee Agency (MIGA)

The IBRD usually works with credit-worthy poor countries, as well as with middle-income governments. Part of the IBRD service is to provide loans, guarantees, analysis, and advisory services.

The IDA provides grants and interest-free credits by working with a number of governments in underdeveloped countries (around 81 countries).

The IFC provides loans on a long-term basis, as well as advisory services, risk management, and guarantees. The bank usually works with the private sector in the form of investment in the poorest countries.

The MIGA provides insurance (e.g., political risk insurance) as a protection related to noncommercial risks for the qualified foreign investors, as well as for the commercial banks, for the purpose of investments in the above countries.

The World Bank provides financial support related to a variety of renewable energy projects around the world. For example, two renewable energy projects in South East Asia to provide electricity for more than 10 million households in Indonesia; and electricity for more than half a million people in Bangladesh (World Bank, Undated). In a similar way, the World Bank has provided financial help in the forms of loans to finance a number of projects, many related to the energy sector in general and rural electrification in particular, such as those projects connected to RE sources.

9.10 Cooperatives organizations

There are a number of descriptions and definitions related to the cooperative work and their objectives. However, a definition by the International Cooperative Alliance defines "Cooperative" as follows (International Co-operative Alliance, 2010):

A co-operative is an autonomous association of persons united voluntarily to meet their common economic, social, and cultural needs and aspirations through a jointly-owned and democratically-controlled enterprise.

A variety of different types of services are provided by cooperative organizations. These services can range from housing, agricultural, health and social, consumer, worker, and financial services. Since this section is focusing on the financial aspects, the financial cooperative service alone will be discussed here. The financial services provided by cooperative organizations are prevalent across the globe, in that the developed and developing countries find this kind of organization a source of important services provider to their communities. Generally speaking, financial cooperatives are locally based and privately owned, that is, to say that the association is the property of the cooperative members. The voting rights are based on membership, instead of the value of the shareholding. Also, as the set-up of the cooperative is based on a group of people working together, the benefit is only to their own members, rather than the customers.

According to the World Bank, financial cooperatives within the developing countries were not always successful. Some of the failure of the financial cooperatives in the developing countries are attributed to political influence, and, according to the above source, cooperatives are used by governments *"for their own purposes."* However, financial cooperatives within the rural areas in the developing countries have managed to operate successfully, regardless of the origin of support they have received, be that governmental or donors support (What is the World Bank, 2010).

Cooperative organizations around the world have large market shares, for example, in Western Europe, there are around 11,000 cooperative banks, with 56,000 branches and more than 33 million memberships (World Bank, 2007). These kinds of figures contrast with the cooperative financial organizations in the developing world, in particular in rural areas where these financial services are badly needed. The structure of cooperatives and their formation are usually born from a lack of interest from investors, both public and private. The need of the local community for a service (or services) has provided the momentum to form cooperative organizations. This is what had exactly happened in different parts of the world. For example, during the first quarter of the 20th Century, when many of the electric utilities were not interested to invest in the countryside to provide power supply for the local population in the USA. As a result, cooperatives were formed for the purpose of providing electricity for every home in the rural areas. Consequently, the "Rural Electrification Administration (Cooperative)" was established in 1935 (Iowa Association of Electric Cooperatives, 2010).

The above organization cooperated with the local population for the purpose of forming "Electric Cooperatives" for a wider plan to provide electricity for the whole of rural USA. Iowa Association of Electric Cooperatives reported that there are around 1000 rural electric cooperatives working in the USA. The same report also indicates that there are at least 39 distribution cooperatives and seven generation and transmission cooperatives in Iowa alone. According to the International Cooperative Alliance, the main working principles of these cooperative organizations are set out in the following seven points:

1. Voluntary and Open Membership
2. Democratic Member Control
3. Member Economic Participation
4. Autonomy and Independence
5. Education, Training, and Information
6. Cooperation among Cooperatives
7. Concern for Community

The structure of a cooperative organization can be summarized in the following points, as shown in Fig. 9.1:

A. Cooperative members elect the Board of Directors, and each member is entitled to stand for election
B. The responsibility for governing the coop is by the board of directors is responsible for governing the coop, which includes setting its direction and hiring senior management (for instance, the Executive Director, CEO, etc.)
C. The Board of Directors is accountable to the membership
D. Executive managers (seniors) are responsible for the day-to-day operation of the Cooperative

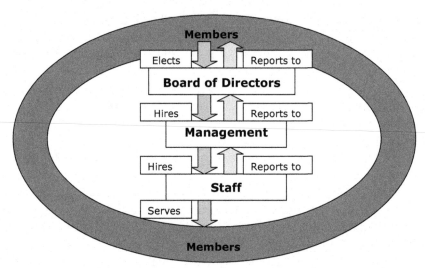

FIGURE 9.1 The structure and operation of a coop. *Redrawn and edited from British Columbia Cooperative Association (2010).*

9.11 Micro-financing

Microfinance started during the 1970s in Bangladesh as a means to provide for the poorer community, that may be excluded by conventional financial institutes, financial services, such as loans and saving facilities (with a minimal appraisal of borrowers and no collateral demand) in addition to advice and consultations.

Microfinance has been established as one way of dealing with poverty alleviation. One of the reasons, which made microfinance a successful financial organization, in particular within the developing countries, is that the established financial institutes can only reach around 25% of the average population, while the rest will find it difficult, if not impossible, to benefit from the services offered by the above institutions, according to a World Bank report dating back to 1995. The basic principle for microfinance is to lend a small loan, which can be repaid over a period of time (usually longer than traditional lending businesses) with an interest rate lower than the local lending market. The money is used mostly for self-employment, to start a small business, to develop an already existing project/business within a local community or it can simply be used for basic household needs.

In regards to Africa, the new microfinance institutions reach only a small number of people, in addition to this, the main products of those members of the microfinance institute are not well suited for short and/or long term agricultural output.

The National Bank for Agriculture and Rural Development provided the following general factors that have contributed to the success of microfinance banking (NABARD, 2010):

1. Microfinance can play an important role in rural areas of the developing countries in that it can directly reach those who are in need of financial services

BOX 9.1 Possible impacts of microfinance on women

....microfinance activities that do not adequately consider the social, cultural and gender impacts of providing access to resources can harm the people that they set out to help. This is a particularly the case in relation to women.

From AusAid, October 2008. Microfinance, Gender and Aid Effectiveness. Australian Government. Fact Sheet. http://www.ode.ausaid.gov.au/publications/documents/microfinance-issues-note.pdf, 22.04.2013.

BOX 9.2 MFI and licensing

To accept deposits from the general public, an organization must have a license. MFIs that are licensed as deposit-taking institutions are generally subject to some form of regulation and supervision by the country's superintendent of banks, the central bank, or other government departments or entities. Adhering to regulatory and supervisory requirements usually imposes additional costs on the MFI (such as reserve requirements).

From The World Bank, 1999. Microfinance Handbook. Ledgerwood. J. https://openknowledge.worldbank.org/ bitstream/handle/10986/12383/18771.pdf?sequence=1. 22.04.2013

2. Each individual microfinance operates in a specific small location, which means a better understanding of the local community needs and their environment
3. Flexible working hours and flexibility in the approach provided by Microfinance leads to wider acceptability among the local community (Boxes 9.1 and 9.2).

The success of microfinance in Asia has not been reflected in the same way as in some parts of the African continent, in spite of the fact that microfinance had been established there. The survey by CGAP has found out that around 79% of microfinance funding in the sub-Saharan Africa originate from donors while only 21% from financial investors (CGAP, 2010, Figs. 9.2 and 9.3).

9.11.1 Microfinance and rural electrification

Electrification projects are able to mobilize Microfinance systems to speed up the electrification process and, as a result, to provide economic, social and a better mechanism in solving a number of negatives aspects connected to rural areas in the developing countries. By providing financial services to the poor, opportunities will be opened in various ways, and the relationship between this type of finance and projects dedicated to rural electrification in the developing countries, should provide the following benefits:

a. Regular income e.g. in the form of self-employment/starting a new business
b. Increasing and/or improving on farming products
c. To increase the likelihood of being able to pay household bills, including electricity bills

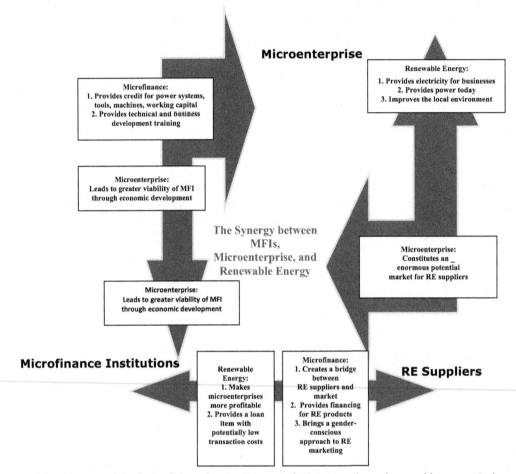

FIGURE 9.2 A representation of microfinance showing it as part of microenterprise and renewable energy. *Redrawn and edited from NREL, Allderdice, A., John H.R., 2000. Renewable Energy for Microenterprise. National Renewable Energy Laboratory Global Transition Consulting, Inc. http://www.nrel.gov/docs/fy01osti/26188.pdf. 9.8.2010.*

d. General economic improvement and growth to the village/town when the local population as a whole has the ability to spend the loans provided by microfinance within their own local market

e. The saving proposed by microfinance for the local population is another way for the basis of future availability of funds and further economic growth

The above financing method will directly and indirectly help the rural electrification projects in that an environment will be created where aspects may not be the same or right for the electrification in the absence of the above scheme. In Bangladesh and parts of India, some of the households use the microfinance finance for the purpose of purchasing SHS (Solar Home System). According to Grameen Shakti, more than 65,000

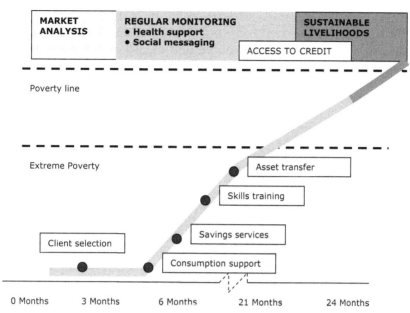

FIGURE 9.3 Microfinance graduation model. *Redrawn from CGAP, 2010. Announces the Results of the 2008 Microfinance Funder Survey. http://www.cgap.org/p/site/c/template.rc/1.26.4501/. 6.9.2010.*

SHS in rural Bangladesh have been bought and installed. The number of installations is made possible by purchasing their systems on microcredit with affordable terms, tailored to individual specific needs. Most of the funding itself for the microcredit system originated from the World Bank, using local companies and businesses to supply the SHS systems.

9.12 Consultative Group to Assist the Poor (CGAP)

The CGAP was established in 1995 with funding of £30 million for a 3-year-period from the World Bank. Housed and associated with the World Bank, the mission for CGAP, as defined in its charter, is to facilitate access to financial services to the poorest sections of the society. As a research center and policymaker aimed at the alleviation of poverty, the CGAP obtains its support from a variety of private organizations and development agencies. Areas of research/services range from market intelligence/solutions to advisory services provided to various institutes and governmental organizations.

According to the above source, the outputs in its priority areas have achieved the following:

Promoting institutional diversity; promoting diverse financial services to a broad range of clients; building financial market infrastructure; fostering sound policy and legal frameworks; improving the effectiveness of microfinance funding; efficiency and governance.

As CGAP supported financial schemes exist in a number of underdeveloped countries, the microfinance system constitutes an important part of its work. The above support helps in various ways to speed-up the local development, including rural electrification projects. The organization's make-up can be summarized as follows:

1. Consultative Group of Member Donors (The governance of CGAP and membership)
2. Executive Committee (donor representatives plus microfinance Executives)
3. CGAP investment Committee
4. CGAP Staff

9.13 Other banking and financial support services

There are many other international financing institutions and banking services that provide financial help directly and/or indirectly to a variety of rural electrification projects, such as the Asian Development Bank(ADB), African Development Bank (AfDB) Group, and Inter-American Development Bank (Box 9.3).

9.14 Asian Development Bank

The ADB Bank is an international development and financial institution established in 1966 on the recommendation of ECAFE (Economic Commission for Asia and the Far East) for the purpose of reducing poverty in the member countries in the Asian and Pacific region. However, the ADB did not start its financial and development activities until 1 January, 1967 (ADB, 2010). The bank is presently made up of 67 members, the majority of them are in Asia, while 19 members are located in different parts of the world. The ADB's work partners are from both the public and the private sectors, such as governmental and nongovernmental departments, development agencies, and local

---

BOX 9.3 Microfinance and women in underdeveloped countries

The Asian Development Bank (ADB) during the evaluation of five microfinance projects, which supports the consideration of gender within the microfinance, has found out that the projects produced positive effects on women lives and, at the same time, indicating that women had:

1. An important role for household cash generation
2. Have a greater role in decision making regarding expenditure decisions and savings
3. An important role in business decision, as well as, they can generate more income working on their own
4. Gaining more skills while at the same time increasing their support system
5. Greater acquisition of assets

BOX 9.4 Do governments do a good job of delivering microcredit?

"…the vast majority government microfinance programs do a poor job of delivering retail credit. Such programs are usually subject to political influence, high default, continuing drain on national treasuries, and sometimes lending based more on the borrowers' influence than their actual qualifications."

From CGAP, 2013. What is Microfinance? http://www.microfinancegateway.org/p/site/m/template.rc/1.26. 12263/. 22.04.2013.

---

community/foundations organizations working for similar aims and objectives. The financial services provided are in the form of loans, grants, technical assistance, and financial advice/consultations.

The structure of ABD can be summarized as follows:

1. Board of governors, with each state represented by one governor and an alternative governor
2. Board of directors (there are 12 regional and nonregional directors)
3. President (usually from Japan) and three deputies
4. Financial support is provided to regional, subregional, and national projects as ADB first priority
5. ADB main resources are generated via subscriptions, interest on undisbursed assets, as well as, from borrowing on the capital markets
6. The major subscribers to ADB are Japan, USA, China, India, Australia, Germany, Indonesia, and the Republic of Korea
7. With highly concessionary terms, the majority of the ADB loans provided from its special funds, such as Asian Development Funds (ADF), are allocated mainly to the poor countries
8. The ADF obtain their funds from ADB developed countries members in the form of voluntary contribution, as well as loans. The largest loan provider usually originate from USA (Boxes 9.4 and 9.5).

9.15 African Development Bank (AfDB) group

According to AfDB, the aim of the group is to help reducing poverty in Africa, as well as to use the continent's resources for the purpose of improving the economic and social life by providing the mean for a higher standard of living for all the people of Africa (AfDB, About Us, 2010).

As a financial development institution, the AfDB is made up of African Development Bank, the original financial institution established during 1963 in Sudan, the African Development Fund, and the Nigeria Trust Fund, all presently under one management

---

BOX 9.5 Microfinance positive approach to business

Focus on the business goals
Set pricing policies and be sure that interest rates cover their costs.
Adapt the funding and support to maturity.
Use cost-sharing mechanisms.
Engage partners in performance-based contracts.
Technical support should have a continuous local presence.
Collaborate with others supporting the same institutions.
Make a long-term commitment.

From CGAP, October 2006. Microfinance Consensus Guidelines: Good Practice Guidelines for Funders of Microfinance. http://www.cgap.org/p/site/c/template.rc/1.9.2746. 4.9.2010.

(AfDB, History, 2010). By the end of 2007, The AfDB had 77 members, of which 53 are African countries members, and the other 24 are nonAfrican countries (AfDB, History, 2010). The AfDB cooperates with a number of international financial institutions, such as the World Bank. In fact, the World Bank signed a Memorandum of Understanding (MOU) in the form of partnership with the AfDB; also, both banks with the cooperation of the International Monetary Fund, established JAI (Joint Africa Institute), which, among other activities, allocated the task to train 400 participants each year in areas related to social and macro-economic issues (World Bank, 2010, AfDB). The decision-making body of the AfDB is the Board of Governors. The Board of Governors is made up of a member (representative member) of each member state. The African member countries have majority ownership of the Bank. The AfDB Bank structure can be summarised in the following points:

1. The President - Elected by the institution's Board of Governors for a five year term;
2. Board of Governors - Apart from electing the AfDB president, they decide on increases to the Bank's capital and responsible for the admission of new members
3. Board of Directors - Decision making concerning projects and policies of the AfDB. Executive Directors are elected for three-year terms, renewable once

In addition to poverty the commitment for the reduction of poverty, supporting environmental issues and the work related to gender inequalities, it has been reported that the AfDB operations cover other areas, such as transport, agriculture, health and education. The AfDB is one of many other development organisations which are active in Africa, however, the loans provided by the AfDB are usually smaller than other loans provided, such as those originated from the World Bank and donors' agencies. According to Weiss, the IfDB '...*the bank has long been considered to be the least capable of the regional development banks*'.

9.16 Energy subsidy

A definition for energy subsidy by the UN Environment Program, Division of Technology, Industry and Economics, states:

An energy subsidy is any government action that influences energy market outcomes by lowering the cost of energy production, raising the price received by energy producers or lowering the price paid by energy consumers.

Having a standard definition for energy subsidy can simplify the approach to the subject itself, however, when it comes to the word "subsidy" itself, branches of large different financial aspects will be open to the reader. As an example, the following subsidy titles are just a few out of many: Indirect subsidies, Labor subsidies, Housing subsidies, Cross subsidy, Infrastructure subsidies, Actionable subsidy, Environmental subsidy, Nonspecific subsidy, Prohibited subsidy, Nonactionable subsidy and so on.

The energy subsidy (or electricity subsidy, as in our case) can be attributed to a number of factors of why and how this kind of financial help (partial or total) could work well in certain situations, while, on the other hand, it may not be beneficial in many other situations. The opinion of improving social efficiency in providing a better standard of living may be in contrast of how effective in solving the energy shortages in the countryside, on a short- and long-term basis. The possibility of higher taxes in order to provide subsidies to firms and businesses may encourage inefficiency, as these enterprises become dependent on government aid, at the same time, the public authority who is providing the subsidy may have poor and/or inaccurate information and data concerning the service being provided by the energy/electricity provider.

Having said that, various governments will argue about the reason of why a subsidy should be provided for a particular domestic service or industry, in that their justification is to promote jobs creations and to protect the local/national industry against foreign competition. In addition, it is a way of protecting the environment by making sure that the business will adhere to the governmental regulations, where the environment is concerned. Also, to encourage rural and urban development for the benefit of all, and to provide a platform where the energy sector can develop in a way, which make energy services affordable, in particular to the poorer section of the community, as well as to the remote areas of the rural parts of the country. Another important justification for subsidy by the government is connected to the "national security" by making subsidy an important move by the government to help in reducing dependency on imported energy (UN Environment Program, 2008).

9.16.1 The positive aspects of subsidy

For rural electrification, subsidies can be vital for the success of a number of projects, where the main aspect of the located area is poverty. For this reason, an electrification project will not be able to succeed without the financial support from the local authority and/or the national government in the form of subsides to help the local population.

With relatively high cost for the rural electricity, compared to the individual family low income and low saving, as well as the low willingness among the population to pay, all provide a general factor that certain rural community will not be able to meet the billing cost for supplying them with electricity. Subsidy is beneficial in the above case; within a specified period of time, until the local economy establishes itself and, as a consequence, the annual earnings for the average family will rise.

The subsidies will be in the form of a capital investment, possibly within a successful project, where the dividend earned will be allocated to the power generation utility for the purpose of lowering the electricity cost to individual customers.

9.16.2   The negative aspects of subsidy

According to UN Environment Programme, many types of the energy subsidy are not helpful to the goal of sustainable development, in that:

A. Increase the levels of energy consumption and consequently, higher amount of waste affecting the environment
B. Burden on the government budget that could have other major negative aspects, on the society as a whole
C. Weaken the private and public sectors investment of the energy sector
D. Not reach the people who need the subsidy most, even though the subsidy is meant to help the poorest section of society

9.16.3   IEA global subsidy

By looking at the prediction IEA made more than ten years ago - published in 2010 - the global subsidy related to fossil fuels did not change that much for more than a decade ago, in fact, in certain cases, the opposite is true. The survey that carried out by IEA to find out about 37 countries which offer subsidies for the purpose of reducing fossil fuel prices, that lead consequently to a higher consumption of fossil fuels, which is the issues of subsidies still similar in 2020, concluded in 2010 the following:

1. Fossil fuel consumption subsidies amounted to $557 bn in 2008, an increase of $215 bn compared to 2007
2. In 2008, the highest subsidies were in Iran, around $101 billion, or, as the IAE put it, close to a third of the country's annual central budget
3. By gradually reducing and eventually removing subsidies, this will provide "an incentive" for efficiency for everyday use of energy, and encourage the use of other types of fuels that emit lower greenhouse gaseous

The phasing out of subsidies, according to IEA, between 2011 and 2020 can be achieved as follows:

A. Reducing global energy demand by 5.8% by 2020 (not met)
B. Reducing global oil demand by 6.5 mb/d in 2020 (not met)

C. The reduction of CO_2 emissions by 6.9% by 2020 (not met)

D. Copenhagen Accord implementation and phasing out subsidies as a *"complementary steps toward achieving the 450 Scenario"*

E. Careful considerations should be made related to the regulations of phasing-out subsidies in conjunction with kerosene, LPG, and electricity

F. It is important to make available all information and data about the availability and transparency of subsidy related to energy in order to begin global subsidy reform, as pointed out by the G-20 group of countries

This is a global general approach by the IEA as a way to reduce subsidy mainly on fossil fuels, but it has not occurred the way IEA predicted it. With subsidy related to rural areas, particularly where there are the needs for electrification projects, the IEA approach should differ considerably in that electrification projects are simply another way to reduce dependency on fossil fuels, particulary where aspects of RE supported and applied.

9.17 Rural electrification financial challenges

Challenges facing rural electrification in the developing countries are a big subject, which deserves to be looked at in detail separately from all the other topics, including topics related to the financial aspects. However, as a way of emphasizing the importance of the finance in any project, in particular electrification projects, the following summarized points are a reminder of how challenging the task can be when it comes to the actual financial aspects of a project:

1. Local institutions may not be able or willing to provide the finance needed
2. In many cases, even if there is the availability of finance, the likelihood of inadequate financing could delay the work, which consequently that in itself may increase the overall final cost
3. Even when the financial estimate over the lifetime of a project is close to the actual figures needed, the underestimation over the next five years to sustain the work on the project (such as money provided in the form of subsidies), the figures in the majority of cases fail to reach the actual cost. In addition to the above, especially in developing countries, government fail to live up to their promises in some cases
4. Lack of regular electricity supply to the local community is a deterrent to investment, and consequently, to further development and further financing
5. Selling of power to the main grid by small power investors should be a straight forward contract, and governmental bureaucracy, as well as the imposed unnecessary regularity delay, should be speeded up or avoided
6. Governmental funding sources should be looked at, and modified if necessary, and diversified in order to match with the requirements of the planned Government schemes and rural electrification projects

7. Subsidy for successful rural electrification projects should be gradually decreased with the incentive of lower or no taxation for future development and expansion from the private sectors
8. Cooperatives, microfinance, and other forms of finance should establish themselves long before an electrification project is being considered
9. It is very well known that one of the major problems for rural electrification in the developing countries is the lack of finance and poverty, in general. However, whenever there are plans to set up cost-recovering prices, such as deposit or insurance, in the majority of cases, they are resisted by the users
10. In order to increase efficiency and reduce bureaucracy, privatization should be encouraged as part of attracting private investment to the rural areas
11. Investment in the developing countries power sector has to compete with the cycle of large re-investment in the EU, as well as in other countries
12. A higher standard of living and aging in emerging industrialized countries such as China and Korea, as an example, can reduce saving and consequently may contribute to *capital shortage,*, i.e., most of the projects will not be able to attract investment unless they are best/exceptional in their approach
13. Rural utilities should be registered as companies under national commercial law
14. The utilities should be independent in their work and, consequently, national and local government should not interfere in their operations
15. There should be a good approach to the utilities accounting procedures and presentation, and therefore, should reflect the actual financial company state in order to prevent regulatory and political risk and intervention
16. For the purpose of attracting investment, electricity tariffs, as a way of recovering cost, should be introduced
17. Prices should be at marginal cost; however, the subsidy should be introduced only when there is a genuine cause related to affordability
18. Cross subsidy should not be implemented
19. Competition should be encouraged and introduced via IPPs (Independent Power Producers)
20. The size of the proposed power company should be considered in the understanding that smaller electricity providers may not be able to survive as a businessin the long term, compared to a larger firm. Reportedly, an efficient electricity provider within a developing country may have around two to three million customers
21. Only large power companies should consider horizontal and vertical separation
22. For achieving higher economic success, rural electrification should be set up as a separate organization
23. The social activities of the company should be separate from the commercial one

9.18 Conclusion

By fully understanding the mechanism of the financial markets from various approaches, i.e., internationally, nationally and locally, for the purpose of long term investment in rural electrification, the overall view of financing a local community project will be a simpler task. The conclusion from this chapter should be taken with the view that rural electrification, in essence, should be a private enterprise, even if the funds and planning have originated from governmental or public sources. As it is well known, private enterprises such as power stations in the rural areas, automatically come under the local and national regulations of the country concerned. Therefore, the direct governmental intervention of local electrification projects would be unnecessary, to say the least, and unhelpful, where bureaucracy, delay, and corruption are the norm in a number of developing countries. Competition is a vital aspect of improving the service and reducing prices. This important factor, i.e., competition, can only surface where genuine privatizations are taking place.

The example set up in some European countries, such as in the UK, where a "watchdog" committee constantly keeps an eye on those private companies providing essential services to the community, is one way of ensuring a better outcome for the customers, as well as for the investors.

Sections of this chapter dealt with the project structures in order to inform discussion of the project aspects and project stages. This kind of approach is a way of examining the general basic outline of a project with the provision of possible solutions, before commencing on the main issue of the financial aspects for rural electrifications. International, continental, and national organizations and their procedures, whenever they are applicable to the rural electrification financing aspects, have been looked at as a possible source of finance for rural electrification. Out of all the organizations examined, microfinance was the most applicable approach for South Asia, as it can be one step toward helping further development of rural electrification projects. The cooperative approach can also be a vital part of the construction and organization of various rural electrification projects across the globe. In fact, cooperative funding differs from microfinance in that it can take direct action in forming a company by involving the local community as members of the organization, whether for the purpose of rural electrification or for providing other types of services. Best examples for rural electrification via cooperative organizations can be found in the West. For example, the speed of rural electrification during the early part of the 20th Century in the USA provides many important lessons for developing countries. These lessons originate from the local people working together, helping to speed up the process in order to provide the service(s) needed for the local community. This kind of enterprise approach usually takes place when public and/or private organizations (as well as individual investors), are/were unwilling or hesitant towards this kind of investment.

References

ADB, 2010. About Us. http://www.adb.org/About/, 8.9.2010.

AfDB, 2010. About Us. http://www.afdb.org/en/about-us/, 9.9.2010.

AfDB, 2010. History. http://www.afdb.org/en/about-us/history/, 9.9.2010.

British Columbia Co-operative Association, 2010. About Co ops. Co-op Structure: Building a Co-operative Economy. http://bcca.coop/content/co-op-structure, 10.9.2010.

CGAP, October 2006. Microfinance Consensus Guidelines: Good Practice Guidelines for Funders of Microfinance. http://www.cgap.org/p/site/c/template.rc/1.9.2746. 4.9.2010.

CGAP, 2010. Announces the Results of the 2008 Microfinance Funder Survey. http://www.cgap.org/p/site/c/template.rc/1.26.4501/. 6.9.2010.

CGAP, 2013. What is Microfinance? http://www.microfinancegateway.org/p/site/m/template.rc/1.26.12263/. 22.4.2013.

AusAid, October 2008. Microfinance, Gender and Aid Effectiveness. Australian Government. Fact Sheet. http://www.ode.ausaid.gov.au/publications/documents/microfinance-issues-note.pdf. 22.04.2013.

ELC − Electroconsult (Undated). Rural Electrification. http://www.elc-electroconsult.com/%5Cdownload%5CEnglish%5CBrochure%5CRural_Electrification.PDF. 16.8.2010.

Environment and Energy, UNDP, 2010. Sustainable energy. Energy for Sustainable Development: Overview. http://www.undp.org/energy/. 6.9.2010.

Global Policy forum, 2010. UN Finance. http://globalpolicy.org/un-finance.html, 17.12.2010.

IEA, 2004. Sources of financing for PV-Based Rural Electrification in Developing Countries. Photovoltaic Power Systems Programme. Report IEA-PVPS T9-08:2004. http://www.iea-pvps.org/products/download/rep9_08.pdf. 20.12.2010.

IEA, 2010. Energy Subsidies: Getting the Prices Right. http://www.iea.org/files/energy_subsidies.pdf. 13.9.2010.

International Co-operative Alliance, 2010. What is Co-operative? http://www.ica.coop/coop/index.html, 10.9.2010.

Iowa Association of Electric Cooperatives, 2010. Frequently Asked Questions - Organizational Issues - Why Do We Have Electric Cooperatives? http://www.iowarec.org/about_us/frequently_asked_questions/organizational_issues. 12.9.2010.

Kalra, P.K., Shekhar, R., Shrivastava, V.K., 2007. Electrification and Bio-Energy Options in Rural India. Part 1, Rural Electrification. http://www.iitk.ac.in/3inetwork/html/reports/IIR2007/06-Electrification.pdf. 2.1.2011.

Leno V. (Undated) New Approaches in Rural Electrification Fund Design for Sustainable Development - the New Paradigm.

Liming, H., 2009. Financing rural renewable energy: a comparison between China and India. Renew. Sustain. Energy Rev. 13, 1096−1103.

MME, 2009. Programa Nacional de Universalização do Acesso e Uso da Energia Elétrica − Manual de Projetos Especiais. Anexo à Portaria No 60, de 12 de fevereiro de 2009. Eletrobrás/Ministry of Mines and Energy, Brasília.

NABARD, 2010. Micro Finance Institutions − Chapter 8. India. http://www.nabard.org/pdf/report_financial/chap_viii.pdf. 27.8.2010.

NREL, Allderdice, A., John, H.R., 2000. Renewable Energy for Microenterprise. National Renewable Energy Laboratory Global Transition Consulting, Inc. http://www.nrel.gov/docs/fy01osti/26188.pdf. 9.8.2010.

Parliamentary Office of Science and Technology, 2002. Access to Energy in the Developing Countries. Postnote number 191.

The World Bank, 1999. In: Ledgerwood, J. (Ed.), Microfinance Handbook. https://openknowledge. worldbank.org/bitstream/handle/10986/12383/18771.pdf?sequence=1. 22.04.2013.

UNEP, 2008. Public Finance Mechanisms to Mobilise Investment in Climate Change Mitigation. http:// www.sefi.unep.org/fileadmin/media/sefi/docs/UNEP Public_FinanceReport.pdf. 12.12.2010.

World Bank (Undated). The Welfare Impact of rural electrification - Chapter Two; World Bank Lending for Rural Electrification. http://siteresources.worldbank.org/EXTRURELECT/Resources/chap2.pdf. 2.12.2010.

World Bank, 2007. Cooperative Banks in Europe. http://info.worldbank.org/etools/docs/library/239521/ Guider_EACB_LandscapeCoopFin-SPEECH.pdf. 4.12.2010.

World Bank, 2010. About Us: What is the World Bank. http://web.worldbank.org/WBSITE/EXTERNAL/ EXTABOUTUS/0,,contentMDK:20040558∼menuPK:34559∼pagePK:34542∼piPK: 36600,00.html, 7.9.2010.

10 ▪▪▪

Cooperative model

Rural electrification has a better rating success if and when the local community participates actively in the proposed project. Crystallizing the local community involvement, in the form of a cooperative model, can be the ideal approach, especially when poverty is the main obstacle for a village/town development.

In this chapter, the electricity supply for the local community using a renewable energy source will be examined and discussed. Electricity generated, such as from solar energy, is presented in the form of a cooperative model for an example town (fictitious) presumed to be allocated in the Indian state of Madhya Pradesh (MP), backed up with a scoring value factors methodology for the purpose of establishing a suitable power system for the proposed town.

For providing a basic approach, which may help in solving some of the problems related to rural electrification, a novel local community action will be needed and could provide a suitable solution. This could mean that the electricity supply provided via a cooperative enterprise model, where the local population can participate in running a solar PV power station, as a first step in solving some of the basic domestic energy needs.

10.1 Example

10.1.1 The model

In order to start the process for a model, obviously, accurate data are needed. A guideline constructed sample of a village or a town will provide good illustrations in establishing an RE system that can be replicated anywhere else in the world. The sample should be able to provide a list of the number of households, where electricity supply is still lacking within the local community. The example below resembles a town in India, which can be from Madhya Pradesh state.

If the assumption made that 44% of the rural areas of India have access to electricity, it means there is still a large number of people living without access to electricity. The problem of estimating the percentage of people who have access to electricity stems from the way in which a village provided with electricity supply is defined. The Census of India defines it in this way: "*A village will be deemed to be electrified if electricity is used in the inhabited locality within the revenue boundary of the village for any purpose, whatsoever.*" However, changes have been made to the above definition by clarifying the basic infrastructure of the supply distribution, the supply to public places and the number of households with electricity supply, which should be a minimum of 10% of the total number of households in a village/town (Coninck et al., 2005, Table 10.1).

Rural Electrification. https://doi.org/10.1016/B978-0-12-822403-8.00010-2

Table 10.1 Hypothetical example for a household source of lighting, such as in an Indian town (Data from Census of India 2001).

Source of lighting	Total	%	Rural	%
Total	10,919,653	100	8,124,795	100
Electricity	7,641,993	70	5,063,424	62.3
Kerosene	3,224,055	29.5	3,024,423	37.2
Solar energy	15,130	0.1	10,535	0.1
Other oil	8,715	0.1	6,762	0.1
Any other	9,638	0.1	6,510	0.1

The idea of locating a village, e.g., in the state of Madhya Pradesh, where a program of electrification is needed, a model of local community cooperative enterprise can be one of the best solutions.

A cooperative locally based model. The model itself should be able to provide all the energy needs, as well as making it possible to apply the same principles later on elsewhere in other parts of rural India, for a similar purpose. The question may arise of why choose the state of Madhya Pradesh, rather than any other state within the national boundary of India? Apart from the abundances of natural resources within the state of Madhya Pradesh, which make it an ideal place to start a natural, sustainable energy system, there are a number of reasons why an electrification project, such as this one, should be located there:

1. Assuming there are 52,074 villages in the state of MP, and let us say that, 4949 of them do not have access to electricity, which may mean that there are around 3036 villages, which have less than 10% access to electricity
2. Applying the definition for an electrified village (see above) may means that there are 7985 villages in the state without electricity, i.e., 15.33% of the total villages in MP. The highest number of nonelectrified villages is located in the Vindhya Region, accounting for 33.46% of the overall nonelectrified villages in MP

10.1.2   Rural electricity consumption and tariff

The consumed electricity in rural India is connected to four categories of end users. These are domestic, commercial, industrial, and agricultural use, such as for the use of irrigation pumps. The cost for the consumption of electricity is usually measured by the supplied meter per kWh usage; however, agricultural usage does not have meter measurement for the amount of electricity that has been used; instead, there is a flat payment rate (National Rural Electric Cooperative Association, 2000). As an example of rural India electricity distribution and consumption, metered end users account for a small proportion of the overall consumption, the rest, i.e., technical, agricultural, electricity distribution losses, plus electricity theft, will make up the overall final balance. The tariff charges also differ in that domestic usage has been allocated lower charges than the rest of the consumers (Table 10.2).

Table 10.2 Examples of consumer annual kWh consumption for rural India.

Consumer	Sanctioned connections	Annual kWh	Annual kWh/connection
Domestic	1166	183,940	157
Commercial	12	1,300	108
Industrial	25	13,320	533
Irrigation pumps	560	?	—

Redrawn and partly adapted from the source: Bharadwaj, A., Tongia, R., 2004. Distributed Power Generation: Rural India — A Case Study. http://wpweb2.tepper.cmu.edu/ceic/pdfs/CEIC_03_07.pdf 1.11.2010.

As illustrated in Table 10.2, the annual energy cost for irrigation pumps has been left without a figure. This is simply because there is no available data for their usage, as they are not metered. The question is how the power supply utility calculates the electricity cost, which they should charge? According to Bharadwaj and Tongia (2004), the cost has been estimated using the following simple equation:

Total kWh (substation) = kWh (metered consumers) + kWh (unmetered consumers) + Technical Distribution Losses + Theft.

There are two known values in the equation, i.e., 1 and 2; however, there are three other variables with unknown values, and these are 3, 4, and 5. For finding out an estimated value of one of the above variables, the estimated figures for the other two variables should be known. The authors cite an example of how the utility normally estimate the overall cost of the power used in employing these agriculture pumps. They estimate the annual cost (kWh) by checking "a few predominantly agricultural feeders." At the same time, the power utilities estimate the technical losses based on data obtained from a number of feeders. The above method cannot provide actual figures, and therefore, it is prone to over/underestimate the overall cost, in regard to the substation and consumers.

10.2 Energy cooperative

It has been reported that more than 70% of the population of India is located in villages where the lack of electricity and other essential services hinders economic and social development. A village cooperative model, therefore, to provide basic and essential services could be one of the solutions for the local community. The majority of the cooperatives around the world have a common principle in that the electricity supply should be provided on the basis of "universal coverage," i.e., it should be provided for all the local people, with no exception, where the cooperative power enterprise operates. This kind of cooperative approach usually adhered to in the Western world has been noticed (to a certain level) in some parts of India, i.e., a number of households connected via a cooperative enterprise have a much higher rate of connection, i.e., four times more than the connection provided via SEB (NRECA (International, Ltd), 2002).

A cooperative model can be in the form of direct benefits to the members or simply to function on a nonprofit basis, i.e., for the sake of providing energy services. The second approach can focus on the quality and re-investment as part of the main objective of the model, while the first one usually concentrates on the maximization of the profit to be distributed to their members, similar to the function of a company with shareholders. Regardless of the choice, a number of factors should be discussed and agreed upon during the initial stages of consultation. These may involve the following points (Boyd, 2003):

A. The type of ownership
B. Membership concept
C. The control mechanism of the community business
D. The size of the business
E. Internal and external relations
F. The social activities

An agreement concerning the above points should provide the platform for the next step in forming a "cooperative power generation" enterprise. The project proposed the building of a solar plant in a village/town where there is no other power supply presently available. However, the village/town model should have the basic infrastructure for present and future economic development, i.e., a positive outlook for agricultural and/or local industries, as well as easy access to main roads.

10.3 Memberships

The first question to be raised is the population size of the village/town, i.e., the maximum and minimum range of the inhabitants within the proposed model location. According to Boyd (2003), in order to have genuine participation and cooperation for the purpose of achieving the aims and objectives of a community enterprise, i.e., fully active memberships, the number of the memberships should be limited. In other words, the enterprise should not be structured as an organization open for unlimited members. This may suggest that a village or a town of around 5000 inhabitants can be ideal for the proposed energy cooperative model. However, this kind of approach should not restrict or limit the considerations for other types of villages or towns, regardless of the population size, in that, an experimental model should focus on the final possible positive outcome, which can be achieved successfully with limited resources. At the same time, the whole focus should be on the benefit of the whole local population, rather than one certain section of the community.

10.3.1   Statistics

According to Census program results for India during the years 2001 and 2011, the following locations from class I to class VI, in the state of Madhya Pardesh, can be presented as an example, i.e., before deciding on the size of the town/village for the cooperative model. A. Bhind (M) [Population: 153,752, Class - I]; B. Ashoknagar (M) [Population: 57,705, Class - II]; C. Chitrakoot (NP) [Population: 22,279, Class - III];

D. Gurh (NP) [Population: 12,450, Class - IV]; E. Hatod (NP) [Population: 9,028, Class - V];
F. Sethia (or Sethiya) (CT) [Population: 4,559, Class - VI]. According to the above source,
the population size-class has been arranged as follows: Class I: 100,000 and above; Class
II: 50,000 to 99,999; Class III: 20,000 to 49,999; Class IV: 10,000 to 19,999; Class V: 5000 to
9999 and Class VI: Less than 5000 persons. Abbreviations used above: M — Municipality,
N.P. - Nagar Panchayat (a form of an urban body in India comparable to a municipality),
C.T. - Census Town In India, a Census Town is one which has:

a. A minimum population of 5000
b. At least 7% of the male working population engaged in nonagricultural pursuits
c. A density of population of at least 400 persons per km^2.

10.3.2   The model

The village/town model will be chosen on the basis of the following points:

1. Lack of electricity supply
2. The population size
3. The willingness of the local population to get involved in the management, operation, and maintenance aspects
4. The town/village is located within the administered part of Madhya Pradesh state
5. The availability of recent data concerning the village/town

However, as this will be the first cooperative developed model for the project, the
preference will be given to a location in the form of a village rather than a town, if this is
possible (in our example model a town will be considered). This is due to the cost,
timescale, and the final results needed within the lifespan of the project. As mentioned
previously, the proposed cooperative model's main objective is the provision of energy
to the local community. In order to achieve the above structure successfully, a number of
criteria and conditions need to be fulfilled (Holland et al., Undated). These conditions
are summarized below:

1. The size of the village
2. Access to good roads
3. Distance to the distribution network and/or villages nearby already being supplied with electricity
4. Number of consumers
5. Number of established businesses
6. Number of rural industries
7. Comparison of other similar work done in other countries to the new location
8. Availability of public facilities

The project will be established and will function under the following summarized
guidelines:

A. Local ownership
B. Community participation

C. Finding the most suitable solar model related to the geographical location, operation, and maintenance
D. Training those who are able to run the day-to-day operation successfully, and maintenance of the power generator;
E. Devising method(s) for revenue collection
F. Tariff agreement (if possible) with the local/national authority, possibly the rate is in conjunction proportionate to the motor/generator ratings
G. Establishing town/village committee to deal with important aspects of present and future development and for major problems and final decisions

It is not difficult to locate a local community within the Madhya Pradesh state with the above requirements. However, the fulfillment of all the requirements mentioned above depends mostly on the timescale given during the initial stage of implementation. There are other aspects of village electrification, i.e., the village model will be considered under the following additional criteria (Andreas, 2006):

1. The possible influence of the geographic endowment on the electrification
2. The possible influence of the State Electricity Boards (SEB)
3. The influence of the state's general development and structure
4. Possible influence of the main grid, i.e., T&D losses, condition of grid infrastructure and future development
5. A rural village could have a higher priority in states that depend on the agriculture sector
6. There could be a negative effect in the form of higher cost for areas with a large number of villages, i.e., the higher cost will be born out of longer transmission lines and interconnection
7. There could be a positive effect on rural electrification from a large agricultural area, i.e., where irrigation, crop processing, and storage are part of the working village environment where electricity demand will be high
8. It is possible that because of the existence of electrification within regions adjacent to a village, electrification of the village would be part of the future electrification by the state and/or SEB plan if electricity is not connected yet

10.3.3 Town profile

For the purpose of illustration, an example town, referred to "H," in the district of Madhya Pradesh state (hypothetical), has been chosen for the purpose of solar-PV project implementation. Indicative data have been selected regarding this example town (Table 10.3). There are around 9030 inhabitants. The population of the town comprises 51% male (4648) and 49% female (4382). Around 16% of the total population are under the age of 6 (Table 10.3). Although electricity supply has been indicated to be available, reportedly, not all households have access to it. From the information provided regarding the town, access roads, drinking water facilities, population size, and basic town income, all have fulfilled the specification outlined in the Model section provided above, i.e., it can be a suitable community example for the supply of electricity, e.g., via the solar-PV system.

Table 10.3 "H" Town data (Data from Census of India 2001).

Town	"H"	Total population	9030 persons
State	Madhya Pradesh	Females (4382)	1727 (literates)
Country	India	Drinking water facilities	Available
Approach paved roads	Available	Electricity for domestic use	Available
Approach mud roads	Available	Electricity for agricultural use	Available

10.3.4 Town energy data

In order to specify the type of solar power system for the example town, some basic initial statistical data will be needed. These data can range from the overall average power that will be needed per household, and the overall power the system will be able to generate per day. In addition to the above, businesses, industrial, and agricultural will be counted within the supply of the electricity generated from the proposed PV system. These have been included in the calculation.

The usage of power calculation also included streetlights, schools, and places of worship. Below are estimated figures for the proposal of a hybrid solar energy power plant.

All the figures have been estimated in accordance with the size of the population, considering that there are 5 (or for calculation purpose 5.5, in order to get closer to the actual figures) persons living in an average household. Estimated figures are as follows:

Estimated power to be consumed	1.1 MW
Peak load	8 KV
Number of household power connections	1640
Number of commercial power connections	40
Number of Industrial power connections	14
Number of agricultural connections	12
Number of street lights	28
Schools and places of worship	14

Assuming there are 9028 persons who live within "H" town, i.e., 9028/5.5−1640 approximate number of houses in the above town. A randomly selected actual village, such as Bauchhar village, which is located in Gotegaon Tehsil of Narsimhapur district, Madhya Pradesh, with a population of 4179 persons (India Census, 2011). It has been estimated that the average number of people in each household in this village is ∼4. By dividing the population by 4, we obtain the figure of 1044, which is fairly close to the actual number of households in this village (1082).

The following estimated figures have been worked out in a similar way (if compared to Bauchhar village) (Fig. 10.1). Fig. 10.2 provides an illustration for an "average" household power consumption, i.e., with the possibility of combing medium and low loads, at the same time. Some loads can be removed for smaller household loads.

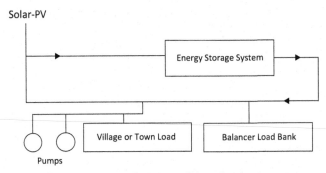

FIGURE 10.1 Basic illustration of the solar-PV system.

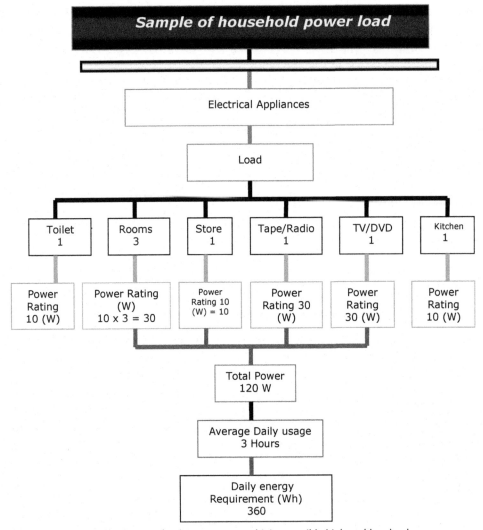

FIGURE 10.2 Power load assessment combining possible high and low loads.

10.4 Financing renewable energy project

The question is how to finance a rural electrification project, particularly a project, such as a solar-PV system, in India? Rural electrification main funding in India can be sourced from:

A. Corporation specialized in rural electrification projects
B. Plan allocation to the states
C. Direct Funding from the government (e.g., in the form of loans or grants)
D. Commercial banks
E. International finance institutions

In regard to point C above, steps have been introduced by the Indian government for a host of financial incentives. Some of these incentives have been summarized in the following points:

1. During the first year of system installation, there will be 100% accelerated depreciation for tax purposes
2. Lower import tariffs for remaining hardware and parts
3. There is a 5-year tax holiday for power generation projects
4. For power generated via the RE system and fed to the grid, the government will provide "remunerative price" under alternate power purchase policy
5. Banking facility plus wheeling of power
6. Subsidies and/or financial incentives for expensive renewable energy devices

10.5 Methodology

The starting point for this work has been based on the review of the literature for a variety of rural electrification projects in different parts of the world, including those specifically designed with the cooperative model in mind. Examples related to cooperative models, where India's rural areas are concerned, have been examined, as well as from those cooperative examples located in different parts of the world, i.e., in relation to the actual suitability and social/cultural aspects. The above study should complement the fieldwork.

10.5.1   Renewable energy analyser two (REA2) approach

A methodology has been designed where the adopted approach is based on what is commonly known as the *bottom-up-approach*, i.e., in the form of building a pyramid (Altawell, 2014). This may mean that all the starting basic factors are mostly known, and therefore, the process starts in recognizing the level of importance and functionality of these individual factors. The building and activation of these factors, in one form or another, will take the methodology to the next level of operation and/or construction.

The methodology, therefore, will consider the construction of, symbolically speaking, a pyramid where the aim and objectives of the methodology will be fulfilled at the top of the completed construction. This may mean that the following steps should be taken in the correct sequence, i.e., as and when each step has been completed separately and/or in relation to other steps (Fig. 10.3).

1. Listing of all known factors on the ground where the proposed cooperative model will be located and function. This will include a number of areas many of which have been mentioned and discussed in the previous section of this paper;

2. Analyzing individual factors on their own merit and the level of importance of each factor in relation to the project. The viability/cost to present and future activity of those local communities involved in running and maintaining the solar energy cooperative enterprise;

3. Compiling all the factors, i.e., hardware, labor/workforce, experts, and energy sources under agreed guidelines, as well as under the principles of a "cooperative model" for the purpose of generating electricity. The overall cost will be examined in the light of the project business viability, present, and future operational aspects and predicted efficiency of the service for the local community;

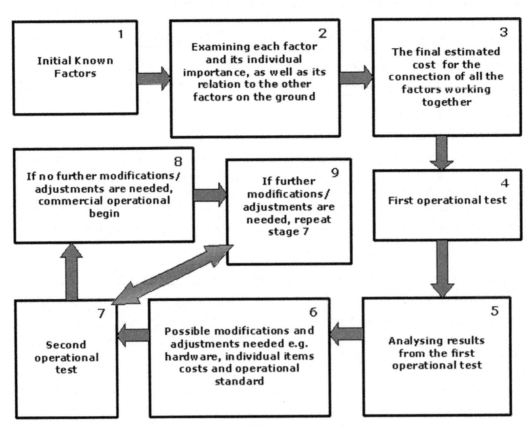

FIGURE 10.3 Methodology approach and stages.

4. A committee of the local entrepreneurs and engineers and other cooperative members will report and comment on the first operational test of the solar-PV system. The report will be in the form of guidelines and a proposed solution(s) on how to overcome any shortages/negative aspects of the project, as well as on how to improve/speed up the process, as a whole;

5. The report received by those overseeing the project as a whole, in particular where financing, engineering and governmental aspects are concerned, will make the final decision on the viability of the work completed so far, and whether to proceed for the next stage or to reconsider the previous one. The results shown at this stage will provide some of the data required for modification and changes, which should lead to an actual improvement during the following stage;

6. The modifications/improvement needed, whether in the form of technical aspects or human resources/local community engagement will be carried out over a period of time, which should lead to an improvised final operational test, i.e., during the second test;

7. The second operational test will be the actual final test, even if the test did not match the expected improvement and/or the service expected. This approach is simply to confirm whether or not at this stage further changes will be needed before the final operational mode has been finally reached, i.e., the starting of commercial service for the cooperative power supply model;

8. If the decision is made to proceed with the supply of electricity to the local community, and no further steps need to be taken for the time being, this stage will lead to the conclusion of the work. If further work is still needed, then stage 7 will be repeated;

9. The final stage (where the overall activity of the project and the form of a cooperative model) will be part of the daily business activity of the local community. Feedback on the progress of the project will be part of the quarterly report produced by the management responsible for the daily operation of the solar-PV station to the cooperative executive committee (Box 10.1).

BOX 10.1 Solar PV output

For establishing how solar PV output would vary, assumptions have had to be made about where the panels are, the angle they are installed at, and which way they face. This note has been based on analysis of the effect of the varying angle of the sun's rays, rather than experimental results.

DECC, 2012

10.5.2   Structure

The methodology will consider the importance of loads, cost/prices, human resources, taxation, grants, and other relevant factors under the following three main headlines. These headlines are the make-up of the three parts of the pyramid. Starting from the base of the pyramid (Altawell, 2014), the main factors are:

1. Regulations/laws and environmental/social
2. Aspects
3. Technical aspects
4. Commercial aspects

The first step is the regulations/laws and environmental/social aspects, which can be subdivided into the following headlines: Land and water issues, possible interference with other utility infrastructure. Pollution and health issues, public and political acceptance, hazard rating, local and national government regulations, weather conditions, tax incentives, regulations related to the use of local resources, incentive and opportunities for private investment.

The second step is the technical aspects of the project, whether in the form of hardware and/or software issues, as well as factors related to the energy source and energy supply. This can be subdivided into the following headings:

The ratio for dependency on fossil fuels and/or other sources of energy, loads demand, electricity output, system life span, electricity payback ratio, quality of service, local skills needed, incremental capacity, and other related infrastructure requirements and development.

The commercial viability of the project, i.e., from an operational point of view, as well as from the influence of the previous factors mentioned above will be forming the final step in completing the pyramid approach. These commercial factors have been summarized in the following points:

The number of customers, affordability for the end users, prices policy, prices for domestic usage and for local industries, seasonal and daily supply policy, profit/losses and shareholders approach and policy, operational and maintenance cost, staff wages, marketing, and present/future development approach.

By using the above methodology structure, a percentage weighing of value to each factor, the level of factor and cost approach will all be considered during each stage of the project structure and valuation, i.e., it will be decided from the data obtained during the implementing process of the project itself. The work of the methodology is simply to provide a scale of measurement, which will be converted into scoring values for each relevant factor, according to the fact on the ground during and after each fieldwork observation and testing (Altawell, 2014).

10.6 Conclusion

The rural community should be encouraged to work together by forming cooperative organizations as a way of establishing enterprises, focusing mainly on the generation of electricity. The cooperative model will work as a symbol for other villages to implement similar procedures to provide access to electricity and consequently improving their daily conditions in their own community.

An outlined summary of a methodology related to the above has been presented as a means for listing all the factors related to a Solar-PV project in the form of a cooperative model. The weighing of these factors, in accordance with their scoring of value and level of access, has made it possible to have a clearer picture of all the aspects connected to regulations, environment, technical and commercial level/values of the proposed solar-PV system.

References

Altawell, N., 2014. The Selection Process of Biomass Materials for the Production of Bio-Fuels and Co-firing. Wiley.

Andreas, K., 2006. Regional Disparities in Electrification of India – Do Geographic Factors Matter? Swiss Federal Institutes of Technology, Centre for Energy Policy and Economic (CEPE). Working Paper No. 51. http://www.cepe.ethz.ch/publications/workingPapers/CEPE_WP51.pdf 4.11.2010.

Bharadwaj, A., Tongia, R., 2004. Distributed Power Generation: Rural India – A Case Study. http://wpweb2.tepper.cmu.edu/ceic/pdfs/CEIC_03_07.pdf, 1.11.2010.

Boyd, G., 2003. Community Enterprise Companies. http://www.caledonia.org.uk/papers/Community%20Enterprise%20Companies.doc, 26.9.2010.

Census of India, 2001. Housing Profile, Madhya Pradesh State. http://censusindia.gov.in/Census_Data_2001/States_at_glance/State_Links/23_mpd.pdf, 22.10.2016.

De Coninck, H.C., Dinesh, K.J., Kets, A., Maithel, S., Mohanty, P., de Vries, H.J., 2005. Providing Electricity to Remote Villages: Implementation Models for Sustainable Electrification of India's Rural Poor. http://www.ecn.nl/docs/library/report/2005/c05037.pdf, 4.10.2010.

DECC, 2012. Solar PV (Electricity Systems) and the National Grid: a Briefing Note for DECC, 26.04.2018. https://www.gov.uk/government/uploads/system/uploads/attachment_data/file/65683/7335-national-grid-solar-pv-briefing-note-for-decc.pdf, (Accessed 26 April 2018).

Holland R., Perera L., Sanchez T., Wilkinson R. (Undated) Decentralised Rural Electrification: The Critical Success Factors. http://practicalaction.org/docs/energy/Rural%20Electrification.PDF, 29.06.2010.

India Census, 2011. Bauchhar Population - Narsimhapur, Madhya Pradesh. www.census2011.co.in/data/village/490505-bauchhar-madhya-pradesh.html, 8.3.2020.

National Rural Electric Cooperative Association, 2000. White Paper on Distributed Generation. http://www.nreca.org/Documents/PublicPolicy/DGWhitepaper.pdf, 2.11.2010.

NRECA (International, Ltd), 2002. Experiences in Cooperative Rural Electrification and Implications for India - Background Paper. Submitted to: Samuel Tumiwa, Asian Development Bank Manila, Philippines. http://www.ocdc.coop/Sector/Energy%20and%20Rural%20Electrification/ExperiencesinCooperativeRuralElecADBRevisedFinal.pdf 21.12.2010.

11 ⊞

Energy economics

The subject of energy economics can provide a socioeconomic analysis of energy markets and the way society produces, consumes, manages, and regulates energy resources. This is a multilayered discipline covering a vast range of topic areas from macro to microeconomics. This chapter focuses on the aspects of energy economics associated with the mechanism of global supply and demand of energy covering the supply structures and the drivers of energy supply from energy extraction to energy conversion, transport, and delivery to end users. It considers market structures, regulatory frameworks, and environmental impacts. It also analyses demand drivers and demand patterns, including changes in consumer trends resulting from technological, regulatory, and general socioeconomic and environmental factors.

This chapter also analyses global energy trends looking at the current and future outlook of energy. The main focus is on supply and demand trends and the drivers of price and investment impacting global energy markets.

The focus on oil prices, environmental objectives, and energy availability has shaped energy markets in recent times. Environmental targets and energy sustainability have taken center stage influencing decision making and energy policies in global energy markets. The global debate on environmental sustainability does not dismiss historical concerns of "peak oil" and scarcity of energy resources, but brings to the fore the contrast between developed and developing countries in terms of access to energy resources and associated social and economic considerations. The sustainability debate has emphasized the view that environmental goals are no longer considered to be an obstruction to economic development but part of the solution.

Environmental costs not incorporated into energy prices are leading to overuse of energy, and this increased awareness of damaging environmental impacts has triggered changes in energy policy, particularly in developed countries of Western Europe. The concerns over energy security and emission reduction targets are informing the energy policy and regulation in these countries, shaping energy landscapes.

11.1 Energy demand in the 20th Century

The energy markets in the 20th Century experienced a wide diversification with consumption of coal declining from 47% of total consumption in 1900 to less than 30% in 2000 (OurWorldinData.org).

In the second half of the 20th Century, oil became the dominant energy source representing nearly half of the overall global energy mix in the early 1970s (BP Energy Outlook, 2018). In the first 2 decades of the 21st Century, oil continues to be dominant but by a smaller margin. Overall, fossil fuels continue to represent over 80% of global energy requirements with oil accounting for 33%, gas for 23%, and coal for 28%.

Global coal consumption appears to have peaked in 2014 with North America and Europe showing a significant decline in both production and consumption. Asia has shown the highest production of coal, with 70% of the world total and the highest consumption worldwide most of the production and consumption deriving from China. Fossil fuel consumption is now experiencing a significant slowdown even in key Asian countries such as China; this is due to the switch to gas and renewable energy in the power sector and fuel efficiencies in the transport and industrial sectors. In contrast, renewable energy experienced the fastest growth since the early years of the 21st Century among all the energy sources. Fig. 11.1 shows the increased growth in renewable energy consumption in the context of the other primary energy sources amidst a very low basis.

According to BP's Statistical Review of World Energy (2019), world primary energy consumption grew at 2.2% average per year from 2000 to 2017, reaching 13,865 million tons of oil equivalent in 2018. World oil consumption (including biofuels) was 4529 million tons in 2018, representing 34% of the world's energy consumption, with growth at an average per year of 1.3% between 2000 and 2018. World natural gas grew at an annual average of 2.8% for the same period reaching 3309 million tons of oil equivalent and representing a 24% share in the energy mix in 2018. Since 2010, natural gas has become the fastest growing among fossil fuels. World coal consumption reached 3772 million tons of oil equivalent in 2018 accounting for 27% of the total energy mix and growing by 2.7% per year during the same period but showing the slowest growth for energy in more recent years (BP, Statistical Review, 2018), as environmental policies in developed countries start to take effect.

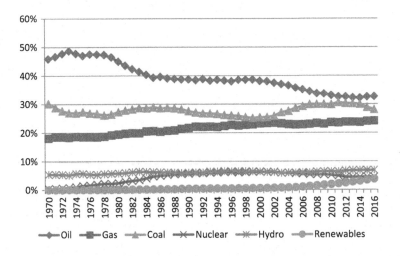

FIGURE 11.1 Energy consumption — share in energy mix (%). *BP Energy EIA, International Energy Outlook, 2017. Total Primary Energy Consumption. Available at: https://www.eia.gov/outlooks/ieo/pdf/0484(2017).pdf. (Accessed 17th July 2018).*

The overall demand for energy is continuing to expand, as increasing prosperity in fast-growing emerging economies lifts millions of people out of poverty and raises incomes. Energy access plays an important role in raising living standards stimulating progress and economic development, helping to mitigate the significant gap between developed and developing countries. Most of the increase in energy demand is continuing to come from developing economies of non-OECD countries. However, the pace of global energy growth is expected to slow down as more focus is directed at using energy more sustainably and achieving energy efficiencies. This has been particularly effective in the transport sector of OECD countries as vehicles become more efficient, and electric cars threaten the hegemony of internal combustion-powered vehicles.

Energy efficiency measures are part of the solution for meeting increasing energy demand with reduced environmental impact. The gains in energy efficiency, together with increasing use of renewable energy, contribute to the reduction of carbon emissions and the move toward attaining international environmental targets.

11.2 Energy transition at the Dawn of the 21st Century

The 21st Century is presiding over significant changes in the way we produce, consume, and manage our energy resources. The energy landscape is changing shape with new centers of oil and gas production reaching global markets, and shifts in demand growth from OECD countries to non-OECD countries. However, the real revolution unfolding in energy markets is linked to the growth of renewable energy propelled by environmental concerns and technological innovation; these are the most defining and irreversible changes heralding a substantial shift in primary energy consumption and decarbonization of the energy mix. Technological advances in renewable energy have lowered production costs and enable competitiveness versus conventional nonrenewable energy.

The world is moving to a new era of clean, safe, and reliable energy. However, we must reset our expectations throughout the "Energy Transition Period" by managing the contributions of both fossil fuel and renewable energy in order to meet the energy demand over the next 20 to 30 years (Box 11.1). While recognizing the importance of renewable energy to meet environmental targets, it is widely recognized that renewables alone are not going to meet the needs of a growing world population; fossil fuels are still expected to represent over 70% in the global energy mix by 2040 with oil and gas accounting for over 60% of the world energy consumption by 2040 (Energy Information Agency, International Energy Outlook, 2017). Oil and gas companies are well aware of the changing energy landscape, and their forecasts are in line with independent energy agencies. BP and ExxonMobil forecast that oil and gas will account for 50% and 60%, respectively, of the energy mix by 2040. It is interesting to note that the discussions around peak oil demand have replaced the "historic" discussion of "Peak Oil Production" - peak oil demand is estimated between 2035 and 2040 according to independent agencies and by oil majors such as BP, Shell, and ExxonMobil.

BOX 11.1 Oil obituary!

Around 60% of the growth in oil demand came from Asia. Although China is the leading global market for the sales of electric cars, it was also the top contributor to oil demand growth, followed by India. Meanwhile, oil demand in the Middle East, a recent source of demand growth, was flat due to oil-to-gas switching in the power sector and efforts to reform oil product prices and phase out subsidies.

While a slowdown in oil demand growth may be likely in the coming years, there are no signs of a peak in demand anytime soon. As noted in the IEA's recent World Energy Outlook (2017) and Oil 2018 reports, it is too soon to write the obituary for oil.

EIA, International Energy Outlook, 2017. Total Primary Energy Consumption. Available at: https://www.eia.gov/outlooks/ieo/pdf/0484(2017).pdf. (Accessed 17th July 2018).

Global renewable energy capacity, including hydropower, has been expanding by 8% in the last 10 years (2007–2018). However, the so-called "new renewables" such as wind and solar power have experienced much higher growth rates globally at 21% and 48% annual growth rate, respectively, for solar and wind power for the period between 2007 and 2018 (Renewable Energy Statistics, IRENA, 2017). But in spite of the growth in renewable energy, fossil fuel will continue to account for the majority of energy demand by 2040.

11.3 Oil demand trends in the early 21st Century

Oil demand in OECD countries has experienced a significant slowdown since the turn of the 21st Century, with a decline of 0.34% versus a 1.6% growth in the last decade of the 20th Century. The decline in oil demand was due to the economic recession, which hit economies of the Western hemisphere particularly hard. However, weak demand in OECD countries also reflects anemic economic growth and further changes in economic activity, the switch to more predominant services-based economies. OECD countries also enjoy significantly higher per capita energy consumption compared to non-OECD, and at the same time, have experienced a reduction in energy intensity by the unit of GDP as illustrated by European Union 28 (Fig. 11.2).

This is the result of the economic changes mentioned above, together with significant fuel efficiency gains in the transport sector. In the USA and in Western Europe, improvements in vehicle fuel economy together with gradual fuel substitution by renewables and switch to electric vehicles will lead to a further decline in fuel consumption in spite of the growth in transportation activity. In addition, high petrol taxation in Europe and a declining population are also contributory factors behind the expected decline in liquid fuel demand in OECD countries. By contrast, the Asia–Pacific region has been enjoying high growth rates in oil demand for several decades. China has

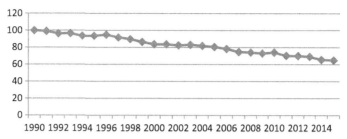

FIGURE 11.2 EU 28 - Energy intensity per unit of GDP (base year 1990). *European Environmental Agency, (2017). EU 28 Energy Intensity Per Unit of GDP. Available at: https://www.eea.europa.eu/data-and-maps/indicators/total-primary-energy-intensity-3/assessment-1. (Accessed 10th August 2018).*

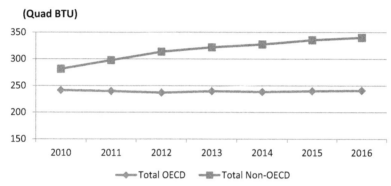

FIGURE 11.3 Total primary energy consumption. *EIA, International Energy Outlook, 2017. Total Primary Energy Consumption. Available at: https://www.eia.gov/outlooks/ieo/pdf/0484(2017).pdf. (Accessed 17th July 2018).*

been leading oil demand growth with an average annual increase of 6.5% for the period between 2000 and 2016, representing an annual average increase in oil consumption of 480,000 barrels per day.

GDP growth in China and other countries of Asia–Pacific outperform all other countries in the world. By contrast, OECD countries have experienced declines in oil demand since 2008 as a consequence of the economic recession, but even prior to the recession, oil demand remained flat. As illustrated in Fig. 11.3, 70% of the increase in global oil consumption derives from non-OECD countries. The combined Asia and Middle East group is expected to contribute to 80% of the increase in global oil consumption. Both regions benefit from high-income growth, while the Middle East benefits from access to ample and relatively inexpensive oil resources.

11.4 Natural gas consumption and the LNG revolution

Natural gas consumption is estimated to experience the fastest demand growth on average in the first half of the 21st Century among fossil fuels. Growth has been greater in non-OECD countries that have been expanding industrial sectors and electricity demand with a 3.7% average growth per year versus 1.3% in OECD countries. Consumption in non-OECD countries is projected to grow an average of 1.9% per year from 2015 to 2040 in contrast to 0.9% per year in OECD countries (EIA, International Energy Outlook, 2017, Reference case).

The electric power sector accounts for nearly 70% of the projected increase in total consumption between 2015 and 2040 (BP, Energy Outlook, 2018). Natural gas-fired generation is attractive for new power plants because of low capital costs and relatively low fuel costs. In the industrial sector, natural gas-intensive industries, such as chemicals, refining, and primary metals, are expected to continue driving industrial demand for gas. New opportunities for LNG in the bunker fuel market as stricter regulation in sulfur content for fuel are pushing some in the shipping industry to switch to LNG — the widening of the spread between oil and gas prices are favoring LNG and making the switch more compelling particularly in the USA (Fig. 11.4).

The strong competitive position versus other fuel sources enhanced by abundant resources is making natural gas a clear favorite for power generation and the industrial sector. Natural gas is becoming the fuel of choice for power generation because of fuel efficiencies and being relatively cleaner compared to coal burning.

The development of vast natural gas resources in many regions of the world and the expansion of LNG facilities creates the conditions for the natural gas industry to thrive. Shale developments in the USA prompted by advances in horizontal drilling and hydraulic fracturing techniques tapped into new technically recoverable reserves triggering an energy revolution. In the past, the US gas market remained isolated and remarkably distinctive from the rest of the global gas market. However, the US shale gas

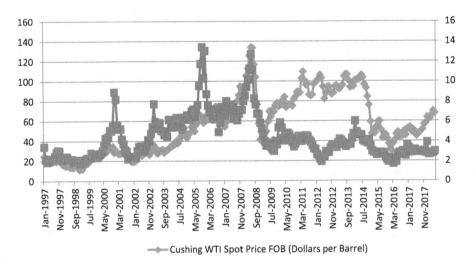

FIGURE 11.4 WTI crude prices versus Henry Hub Natural Gas prices. *EIA, Database, 2018. WTI Crude Prices versus Henry Hub Natural Gas Prices. Available at: https://www.eia.gov/outlooks/aeo/databrowser/. (Accessed 17th July 2018).*

revolution has spurred natural gas price reductions and catapulted the USA into pole position as one of the largest gas producers in the world and one of the world's largest LNG players in the near future. The Henry Gas hub price averaged just under $3 in 2018 versus an average of $7 million per Btu for Japan and under $6 million Btu for Europe in 2018. As US domestic consumers reaped the benefits of low gas prices, US gas producers have been busy building LNG export facilities as the price differentials versus export markets justify the investment.

11.5 Supply developments in the energy sector

Substantial socioeconomic and environmental changes are taking place and shaping global energy production. Energy companies are refocusing their strategies and starting to diversify their portfolio to consider consumer's preferences, as well as government policies and environmental regulations. One of the key objectives is to continue to lower costs and achieve a competitive advantage. However, renewable energy is growing rapidly, challenging the absolute dominance of oil and gas; the share of renewable energy in the energy mix is forecast to double in the next decade. One of the drivers for change is the technology that will be instrumental in speeding the energy transition; batteries and intelligent grids are just some of the key technology elements. At the same time, the slow-down in global economic demand, as well as slow-down in oil, and gas demand in key Asian economies, has led to readjustment in the company's strategies in the oil and gas industry. Demand for cleaner fuels is driving the expansion of the LNG industry and investment in LNG infrastructure is supporting the expansion. In spite of global trends shaping the energy market, there are substantial differences across the world. The problems in the developed world are shifting to issues of clean energy and security of supply, but for developing countries, the issues are very different and quite complex and are often anchored on aspects of basic access to energy.

11.6 Supply developments in oil and gas markets

World oil production expanded to 99.8 million barrels per day (4474 million tons) in 2018, an increase of 20 million barrels from the year 2000 (BP Statistics, 2018). OPEC is determined to maintain its strategy of capturing the market share and remaining dominant in global crude oil markets. The organization holds 77% of proven global reserves of oil and still accounts for 41% of the world oil production, albeit a reduction from a peak of 52% in 1973 (BP Statistics, 2018). The strong output increase from non-OPEC has been unstoppable and has eroded OPEC market prominence. Non-OPEC production is continuing to rise in spite of some declines of production from mature fields of the North Sea and Mexico. The US Gulf of Mexico, West Africa, and Brazil have experienced significant developments in deep-water production, in addition to some substantial ramp-up in Russia production and rise in the oil sands of Alberta, Canada. The US shale revolution has had a particularly strong impact on imposing new dynamics

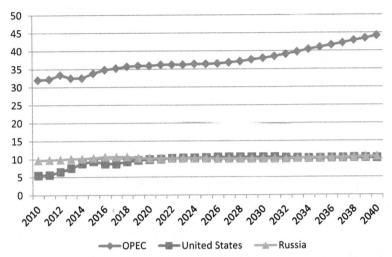

FIGURE 11.5 Major oil producers forecast (million barrels per day). *EIA, International Energy Outlook, 2017. Total Primary Energy Consumption. Available at: https://www.eia.gov/outlooks/ieo/pdf/0484(2017).pdf. (Accessed 17th July 2018).*

in the global oil market. Major contributing factors for the US shale success were the advances in technology in horizontal drilling and hydraulic fracturing which have reduced costs, made shale plays viable, and led to a significant increase in the number of projects being commercially successful in the US, initiating the "shale gas revolution."

Leading the boom is the Bakken field in North Dakota and the Eagle Ford field in Texas, and most of the US shale oil potential and reserves have been estimated at 50 billion barrels. Growth in shale oil production was even more impressive up from a half million bpd in 2011 to well over five million bpd in 2017. This production increase in the USA has had significant repercussions on the global oil and gas market as US imports have dropped dramatically. In the EIA reference case, tight oil production is to reach 8.3 million barrels in the early 2040s, and the total US oil production is expected to maintain the 10 million barrels per day level (IEO, reference case EIA). Outside the US, non-OPEC crude oil production is estimated to grow by 630,000 barrels per day for the period between 2015 and 2040; this increase is mainly from Russia, Brazil, Canada, and Kazakhstan. However, the bulk of the additional supply to satisfy world oil demand is estimated to come from OPEC countries (Fig. 11.5). OPEC is expected to invest in oil capacity in order to maintain its market share within the range of 39%−44% of world oil production.

11.7 LNG supply and trade

LNG global trade has grown by an average of 6% per year over the last 2 decades, with Qatar becoming the largest LNG exporter, and Asia, the main import market (Shell.com). Japan became the leading LNG importer following the Fukushima nuclear accident, which had a very significant impact on the global gas market. As a result, Asian LNG prices shot up to an average of $16/MMBtu, accelerating a trend toward price divergence versus continental Europe. However, 2014 marked a shift in the fortunes of the oil and gas sectors. The collapse in oil prices triggered a fall in gas and LNG prices as the vast majority of LNG projects are long-term oil price linked or oil-indexed contracts. In addition, the restart of some nuclear plants in Japan and generally weaker than expected Asian LNG demand in 2016 and 2017 were factors impacting LNG prices. At low price levels, some LNG projects have struggled to attract contract buyers with some LNG projects under consideration facing delays or even cancellation.

LNG is undergoing an exceptional period of supply growth, which is expected to lead to global oversupply that will impact the LNG market for a few years until demand catches up with installed capacity. Waves of new LNG production are starting to flow into global markets after many years of limited supply. LNG liquefaction capacity is expected to grow at 7% per year until 2020. The markets will be flooded by the production wave, particularly from Australia, followed by the production wave from the USA. At the same time, demand is expected to grow by an average rate of 5% per year until 2030, lagging behind growth in LNG supply until 2020, creating a temporary oversupply of LNG, according to Shell's 2017 LNG report.

New LNG projects will add 124 million tonnes to the global LNG market by 2020; the bulk of the new LNG capacity currently under construction will come from Australia with 40 million tonnes and 60 million tonnes from the USA (Fig. 11.6). Ongoing projects in Malaysia, Russia, and Cameroon will add an extra 24 million tonnes per year of LNG capacity (Box 11.2).

In addition, a large number of LNG projects under consideration in North America could potentially add another 100 million tonnes per year of LNG capacity. However, these additional projects depend on market stabilization and strong price signals. The recent fall in oil prices has dissuaded some buyers from paying LNG capacity and these players are likely to continue to be cautious at least until the additional LNG production is absorbed by the market.

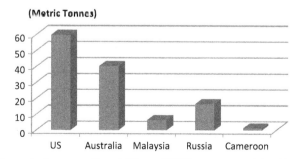

FIGURE 11.6 Main LNG production capacity additions. *GIIGNL, company data, Vygon consulting, n.d. LNG Production Capacity 2015 and 2020. Available at: https://vygon.consulting/upload/iblock/3cd/vygon_consulting_us_lng_2017_en.pdf. (Accessed 1st October 2018).*

BOX 11.2 LNG trade

For large-volume ocean transport, LNG is loaded onto double-hulled ships, which are used for both safety and insulating purposes. Once the ship arrives at the receiving port, LNG is off-loaded into well-insulated storage tanks, and later regasified for entrance into a pipeline distribution network.

LNG can also be shipped in smaller quantities, usually over shorter ocean distances. There is a growing trade in small-scale LNG shipments, which are most commonly made using the same containers used on trucks and in international trade, specially outfitted with cryogenic tanks. Other small-scale LNG activities include "peak-shaver" liquefaction and storage facilities, which can hold gas compactly for when it is needed in local markets in the USA during times of peak demand. LNG is also sometimes imported or exported by truck from this kind of facility.

US Office of Fossil Energy.

World LNG trade is projected to nearly triple from 293 in 2016 to 370 million tonnes by 2020 (Shell, 2019). Australia and North America will be supplying most of the increase in LNG in the next 2 decades (Fig. 11.6). Asia is projected to import a large share of the traded LNG, while Europe is expected to remain largely dependent on Russian pipeline gas. Australia is positioning itself to become the largest exporter of LNG by mid-2020 with new LNG capacity with some projects already online and others expected to be completed by 2020, adding to a total country export capacity of 88 million tonnes per year (Fig. 11.7).

By 2020 North America LNG is projected to compete with Russia for Europe and Asian markets. By the mid-2030s, the United States is projected to become the world's largest LNG exporter and a net oil exporter by the end of that decade (IEA, World Energy Outlook, 2017).

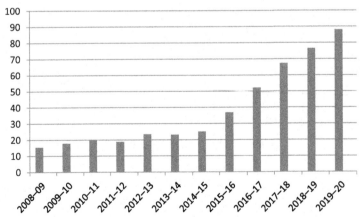

FIGURE 11.7 Australia LNG export capacity (MT/year). *Australian Department of industry, Innovation and Science, n.d. Australia LNG Export Capacity. Available at: https://publications.industry.gov.au/publications/ resourcesandenergyquarterlyjune2018/index.html. (Accessed 30th August 2018).*

11.8 Developments in renewable energy

Renewable energy sources, including hydro, have grown at an annual rate of 2% from 1990 to 2017, higher than any other source of world energy, and representing 15% of world total primary energy supply. There are remarkable differences between OECD and non-OECD on the development of renewable markets. For non-OECD most of the renewables supply comes from solid biofuel and charcoal at 64% with hydroelectricity accounting for 18% of total renewable supplies. These trends are the result of traditional patterns of energy consumption deriving from aspects of affordability and access to available domestic energy resources. However, since 2001 most of the growth of renewable energy for non-OECD Asia derives from the so-called "new renewables," solar and wind power. China accounts for more than 40% of the total additional global investment in renewable capacity.

For OECD countries, most of the growth for renewables energy since 2001 comes from solar power, biofuels, and wind power (Fig. 11.8). Half of the renewable production is applied to heat and power generation. Most of the increase of renewable energy in OECD countries is associated with the implementation of energy policies to encourage the switch from fossil fuel to cleaner energy alternatives. Biofuels in transportation is mandatory in some OECD countries such as OECD Europe.

The reasons for the growing popularity of renewable energy particularly, solar power and wind power can be summarized:

- Technological innovation which enhanced the performance and reliability of solar and wind power
- Progress in energy storage providing solutions for intermittent power from renewables
- Lower capital costs attract investor interest by generating higher investment returns

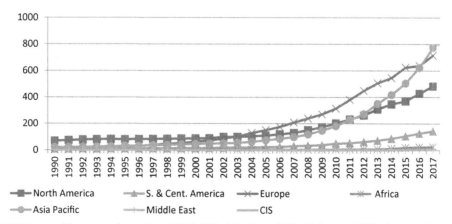

FIGURE 11.8 Renewable energy by region. *BP Statistical Review of World Energy, 2018. Renewable Energy by Region. Available at: https://www.bp.com/en/global/corporate/energy-economics/statistical-review-of-world-energy/downloads.html. (Accessed 10th July 2018).*

- Faster project construction and execution leading to more availability of project financing
- International climate change agreements and commitment toward global reduction of emissions has increased the attraction of investment in clean energy.

Government policy at the national level is a key driver of renewable energy developments. Policymakers are paying particular attention to the power sector, encouraging renewable investment in power projects. At an international level, the Paris Agreement has led the way for the implementation of major reforms in the energy sector aiming to mitigate climate change impacts reducing emissions by switching energy consumption from fossil fuel into an increasing share of renewable energy. Lower costs of energy technologies, particularly solar PV and wind power, in line with the rise of demand for renewable energy supported by government environmental policies, have established the conditions for renewable energy to be competitive versus conventional energy in the power generation sector.

11.9 The future of fossil fuels

The world energy demand is expected to rise by 28% from 2015 to 2040. Most of the increase will derive from non-OECD countries, accounting for more than 60% of global energy increase, with India and China taking the lead accounting for nearly 50% of the global energy growth (IEO, 2017; reference case). By 2040, energy use from non-OECD countries is expected to exceed the entire OECD consumption.

Population and global economic growth are underpinning the expected increase in global energy demand. The global economy is estimated to grow at an average rate of 3% per year for the period between 2015 and 2040; the world population will expand from the current 7.6 billion to 9.2 billion by 2040 (United Nations, World Population Prospects, 2017 version). GDP growth in China and other countries of Asia–Pacific outperforms all other countries in the world (EIA International Outlook, 2017).

In spite of the rapid growth of renewable energy, fossil fuels are still expected to be the dominant source of energy by 2040, representing the majority of the total energy use by 2040 (IEO, 2017). Coal is the only energy source expected to remain flat (EIA, 2017, Reference Case). Coal use for power generation is gradually being replaced by natural gas, renewable energy, and nuclear power. China is the largest consumer of coal, but it is now also slowing down the development of coal projects and investing in renewable energy for power generation. Natural gas will experience a very fast growth among fossil fuels growing at 1.4% per year between 2015 and 2040. The fastest growth, however, will come from renewable energy with an estimated growth of 2.3% per year, while the second-fastest will be nuclear power with an expected increase of 1.5% per year for the same period. However, oil is still expected to be the largest source of world energy consumption for the next 2 decades (EIA, International Energy Outlook, 2017).

Oil demand growth is expected to be quite robust in non-OECD Asia, particularly India and China, in the next 10 to 15 years, but as these developing countries catch up with the more mature world economies, the pace of growth slows considerably. China is the world's second-largest oil consumer, and it has become the largest net importer of oil from 2013. As the Chinese economy develops from a manufacturing basis to a more services-oriented economy, the transportation sector will drive most of the growth in the energy sector (Box 11.3).

Oil consumption growth in non-OECD Asia has been rapidly outpacing production. The Middle East has supplied most of the Asian oil requirements. In 1990, Asia imported 33% of its oil requirements from the Middle East, but by 2010, 48% of oil came from the Middle East. This trend will continue in the future with producers from Russia and Central Asia, making up the difference to meet incremental Asia oil demand (Fig. 11.9).

Oil demand growth in the other non-OECD countries of the Middle East, Africa, and Latin America is expected to be more moderate at a rate of about 1.5%, 0.8%, and 1.3% per year, respectively. Oil demand in the Middle East increases significantly by 3.2 million barrels per day during the forecast period. The growth is driven by strong population growth and higher incomes, but fuel subsidies also play an important role in

---

BOX 11.3 Overall UK renewable electricity generation

Total renewable capacity increased by 14% between 2016 and 2017. Most of the increase derives from wind, accounting for 75% of the capacity addition. In 2017, onshore wind regained the highest share of capacity, and it also held the highest share of generation (at 31.7% and 29%, respectively). The main use of renewable energy is to generate electricity. In 2017, electricity generated from renewables increased by 19% in 2016, from 83.1 to 99.3 TWh. Renewable sources provided 29.3% of the electricity generated in the UK in 2017 compared to 24.5% in 2016, an increase of 4.8 percentage points (measured using the "international basis," i.e., electricity generated from all renewables except nonbiodegradable wastes as a percentage of all electricity generated in the UK). Taken together, onshore and offshore wind represented 79% of the total increase in electricity generation; onshore wind increased by 8.2 TWh (39%) and offshore by 4.5 TWh (27%). This was due to a combination of increased capacity and unusually high wind speeds. The third- and fourth-largest increases in electricity generation (in absolute terms) were plant biomass (1.2 TWh) and solar photovoltaic (1.1 TWh). Landfill gas generation fell by 0.4 TWh, (8.9%) to 4.3 TWh, and cofiring with fossil fuels also fell by 54%. While bioenergy dominates on a fuel input basis, hydroelectricity, wind power, and solar together provide a larger contribution when the output of electricity is being measured.

Digest of UK Energy Statistics (DUKES): Renewable Sources of Energy, n.d. DUKES chapter 6: statistics on energy from renewable sources. IGU Wholesale Gas Price Survey 2017 Edition. Available at: https://www.igu. org/sites/default/files/node-document-field_file/IGU_Wholesale%20Gas%20Price%20Survey%202018% 20Final.pdf. (Accessed 20th October 2018).

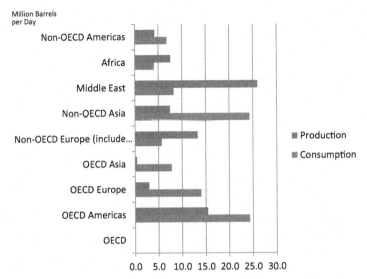

FIGURE 11.9 Oil production and demand by region (2015) (millions of barrels per day). *EIA, International Energy Outlook, 2017. Total Primary Energy Consumption. Available at: https://www.eia.gov/outlooks/ieo/pdf/0484(2017). pdf. (Accessed 17th July 2018).*

the steady increase of liquid fuel. Some reforms in the subsidy regime are expected, and this will slow down the growth rates in this part of the world.

In Russia, the largest economy of the Former Soviet Union, oil demand growth is relatively slow because of significant fuel efficiency improvements achieved in the country's energy-intensive industrial and automobile sector. It is also expected that consumption in the residential and commercial sectors will decline as cuts in fuel subsidies are fully implemented.

11.10 Energy investment trends

Investment decisions in the energy market are influenced by a multitude of factors involving projections of future consumption and production of energy. Questions regarding climate change policies and energy security are crucial decision-making influencers. The assessment of opportunities for consumers, industry, and government are also reliable indicators.

The period between 2014 and 2017 has seen a considerable decline in investment in global oil and gas projects; this was a direct response to the collapse of oil prices. However, there have been significant differences around the globe with a very dynamic US Shale sector contrasting with a slump in oil activity elsewhere. The success experienced by the US shale sector is explained by the considerable cost reductions achieved in shale operations.

If we take the IEA and OPEC analysis, we conclude that for the next 20 years, oil and gas investments will be required to meet around half of the increase in global energy demand. This translates into an additional amount of between 12 and 15 million barrels per day (EIA International Energy Outlook, 2017 and OPEC, World Oil EIAInternational Energy Outlook, 2017) with calls on OPEC production of 10 million additional barrels per day by 2040. For meeting this forecast, a considerable level of investment will be required. OPEC estimates an investment of $10 trillion (World Oil EIAInternational Energy Outlook, 2017) in oil and gas to meet demand growth and offset the decline of mature gas and oil assets; this amount includes $1.5 trillion in downstream investments. Cuts of overall oil and gas investment of 26% and 25% between 2014 and 2016 by most oil and gas companies have seriously undermined the ability of the industry to meet future demand for oil and gas.

Global electricity investment has experienced a slight decline in the period between 2014 and 2017 (Fig. 11.10), which had to do with decline in power generation investment partly offset by a small increase in network spending. Investment in renewable energy has also declined between 2014 and 2017, but this is mainly due to lower capital costs as capacity increased by 50% during the same period.

Investment in the electricity sector has surpassed oil and gas for the first time in 2016 and 2017 (Fig. 11.11). This reflects the weakness of investment in oil and gas, and also demonstrates that the electricity sector is attracting investor interest, particularly state-backed investments (IEA, World Energy Investment, 2017). The coal investment, on the other hand, has been declining as this fossil fuel is increasingly out of favor in power generation due to high carbon content. China, the highest consumer of coal, is post-poning and canceling coal-fired power capacity responding to concerns over the air quality and the effects of emissions on the country's lifestyle standards. However, it is still uncertain if this is a trend for the future in a country with vast coal resources and considerable energy demand growth in the foreseeable future.

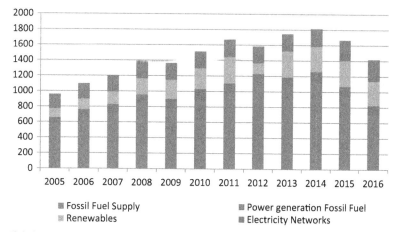

FIGURE 11.10 Global investment in electricity supply (US$ billions). *IEA, Energy Investment, 2017. World Energy Investment 2017. Available at: https://www.iea.org/publications/wei2017/. (Accessed 8th July 2018).*

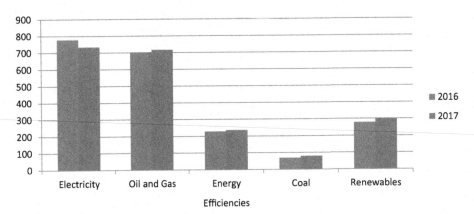

FIGURE 11.11 Global energy investment by sector in 2016 and 2017. *IEA, Global Energy Investment 2018.*

On a regional analysis, China maintains the role of the highest investment destination for energy investments, with 21% of the global energy investment in 2017. China is also the world leader in installed hydro, wind, and solar power. The USA also maintains one of the top investor positions in oil and gas with the focus on shale developments.

Investment in electricity is expected to continue to be strong in spite of a slight decline in investments in 2016 and 2017. Renewable energy is the fastest growing source of power with higher capacity and higher output because of technological advancements and lower unit costs.

However, the absolute value of the capital investment in renewables excluding hydro has been falling since 2015, contrasting with the increase in capacity installation for Wind, solar, biomass, and waste-to-energy, and geothermal. This is mainly due to the substantial decline in capital costs per unit of MW installed, but there has been some slowdown in renewable investment in China and some other emerging markets. However, forecast by the International Energy Agency point to strong investment growth of $7.4 trillion global investment in renewable energy by 2040.

Regarding LNG investment, it has already been mentioned above the remarkable impetus driving investment in the sector. However, the LNG market has become increasingly competitive, and the recent oversupply pressures meant potential suppliers were facing the greatest challenge of finding long-term buyers to secure project funding.

The prevalent business model of most international LNG export projects relies on the offtakers who shoulder all the risk. However, LNG buyers have started to reject this approach and have managed to negotiate more innovative and flexible contracts, which even out the risk exposure along the supply chain; they are more interested in shorter-term and more flexible contracts that enable them to be more competitive in their downstream power and gas markets.

By looking ahead to the long-term future of energy supply, despite the expected reduction in market share for oil and gas, a substantial volume of investments will be required to meet growing oil and gas demand and to offset decline rates in mature oil fields. The lack of upstream investment in recent years is likely to have a serious, detrimental effect leading to a potential squeeze in global oil and gas supply by mid-2020s.

The role of OPEC will continue to be critical to meet demand at favorable low production costs. National oil companies will continue to drive the investment in OPEC countries and continue to grow in influence as suppliers and consumers in rapidly growing markets.

Investment in renewable energy will become a key factor to meet energy demand and to meet domestic and international environmental objectives.

11.11 Oil price drivers

Oil prices are driven by a combination of fundamentals and a multitude of geopolitical factors and economic factors.

Supply and demand are the most important factors influencing price trends. Therefore, falls in global oil prices can be the result of either a significant increase in oil supply or a decrease in oil demand; the latter may be linked with a slowdown in global economic growth with lower oil prices cushioning but not stopping the economic slowdown. Some of the historical events resulted in the withdrawal of supply in the case of the oil shocks of the 1970s or price decline as a consequence of the economic recession, economic slowdown, or market oversupply. The rapid increase in oil prices from 2004 to 2008 (Fig. 11.12) coincided with the strong economic growth and high oil demand growth from the Asia—Pacific region, particularly from China, together with slow supply growth and tight market with little surplus capacity and low inventory levels.

FIGURE 11.12 Brent price–price cycles. *EIA Database, n.d. Brent Spot Price. Available at: https://www.eia.gov/dnav/ pet/hist/LeafHandler.ashx?n=PET&s=RBRTE&f=M. (Accessed 30th August 2018).*

This trend was repeated in the period between 2010 and 2014 except for the supply from non-OPEC countries, which has already been building up since 2010. In addition, during prolonged periods of high prices, consumers tend to focus more on energy conservation and fuel efficiencies—these factors tend to reduce demand in the long term—we have seen this reflected in fuel efficiencies gains in the transport sector. On the supply side, high oil prices trigger investment in upstream oil and gas projects resulting in increased supply, as well as stimulates research and technological innovation. This scenario was well illustrated in the period between 2007 and 2014, with investment in oil projects reaching historically high levels when oil prices reached well above $100 per barrel. The significant increase in production coming from the US shale operations was made possible by technological innovation in shale fracking enabled by high oil prices.

The falling price of crude oil in mid-2014 was largely due to oversupply following a surge in US shale oil production and continued output from the Middle East (despite various conflicts). While the supply levels remained high, demand levels were falling, particularly as certain major economies, such as China and Brazil, had been using less oil. The trigger for the sharp drop in oil price was Saudi Arabia's refusal to cut production in the face of excess oil supply. The collapse in oil prices in second half of 2014 and the persistently low price scenario between 2014 and 2017 has had important repercussions in the oil and gas industry, with oil and gas companies slashing their investment budgets and laying off significant numbers of industry staff.

The price of crude oil started to fall mid-2014, dropping below $30 a barrel (Bloomberg, January 2015). However, low oil prices have not stimulated demand to any great extent in consumer countries as initially expected because of weaknesses in global economies.

The oil industry has been propelled by cycles of boom and bust. Cyclical patterns of high or low price trends are one of the most characteristics features of oil markets. This is rooted in a low level of elasticity derived from the low responsiveness of production and consumption to short-term small price swings. In the oil and gas sector investment cycle is not clearly aligned with supply and demand requirements at any given time because of the lag time between when final investment decisions (FID) are made and projects coming online; it can be up to 5–8 years for conventional oil and gas projects. Investment decisions are also often made based on current or past events rather than on future projections.

The shocks to oil supply or demand can result in large movements in the price of oil over time. It would not take a large shock for oil prices to return to significantly higher levels, and the long lags between oil price changes and the response of oil supply and demand to those changes can lead to cycles in oil prices.

Saudi Arabia is the most influential among OPEC countries with the power to change the direction of oil prices – this has been substantially curtailed in recent years with the emergence of powerful non-OPEC players. Non-OPEC production historically had a cost disadvantage compared to low-cost OPEC production; however, technological innovation enabled continuous gains in productivity and lower costs for shale producers.

There is evidence of non-OPEC oil production, particularly shale oil, having a great influence on the oil price in recent years, as it contributes to significant supply growth but also indirectly by influencing OPEC oil strategies. The market share strategy pursued by Saudi Arabia backfired against the evidence of productivity gains by US shale producers. The market share strategy has been replaced by the traditional market stabilization role.

Inventory levels are also a determinant factor in price movements. The period of 2004−2008 was characterized by tight oil inventories, and this reduced the ability of producers to release oil quickly, adding to existing bullish sentiments. In contrast, the very high inventory levels of the period 2014 to 2017 had the opposite effect.

In mid-2018, oil prices started to climb with Brent strengthening but remaining very volatile within the range of $50 to $80. Between 2014 and 2017, the oil market was "rebalancing" reducing production investment, stimulating demand, and scaling down inventories. The strengthening in oil prices post-2018 is sending a signal to the supply side to increase production in the future.

11.12 Gas prices

Gas prices have been brought down by excess supply and drop in oil prices. This happened because, in many regions of the world, gas prices are still indexed to oil. However, in European markets, a growing share of the gas market is being priced off trading hubs, particularly in north-western Europe while in South-eastern Europe, the price indexation is still the standard way for establishing gas prices. In Asia, the price-indexation is still the way imported gas is traded.

As the natural gas market becomes more integrated globally, prices are expected to become more disconnected from oil prices and more driven by supply and demand fundamentals, with regional differences becoming a closer reflection of transportation costs and currency exchanges. The price of delivering costs of LNG, for example, is likely to reflect shipping costs and liquefaction costs.

In the past, Australia LNG projects have been hit by high costs and collapsing margins. Australia became the most expensive place to develop LNG projects due to high labor costs, difficulties in accessing gas, strong currency, and the recent collapse in oil prices. The collapse in oil prices between 2014 and 2017 placed many projects in jeopardy - some Australian LNG projects are estimated to require contracts of $11/MMBtu for brownfield expansions and $14/MMBtu greenfield export projects. Japan LNG dropped below $5 MM/Btu in April 2016, and European LNG delivery prices hovered around $5.2MM/Btu for the same period (LNG World News, 2016).

However, LNG recovered temporarily to more sustainable levels of $10 MM/Btu in Japan and just over $7 MM/Btu in Europe in 2018 but started to decline again by end of 2018, due to increased volumes coming into the global market from Australia, USA, and Russia (International Gas Union, 2019 Report).

11.13 Global energy by sector: industrial, transport and buildings

Transportation and the industrial sectors are the major driving forces for oil demand as other sectors are more likely to switch to other fuel sources such as gas or coal or renewable energy.

The industrial sector has the largest share of energy consumption by the end-user sector, and accordingly, the EIA will continue to represent over 50% of the total energy consumption in the foreseeable future. Most of the increase will derive from non-OECD countries due to economic development and rapid industrialization. However, compared with other sectors, the industry will experience the slowest growth with 0.7% versus 1% for transportation. Demand from global building sector represents 29% of total energy demand.

More than 50% of global oil consumption is absorbed by the transport sector, but oil has lost market share on both power and industrial sectors increasingly replaced by gas and renewable energy. The growth in the transportation sector in non-OECD countries is driven by economic growth as improved living standards lead to increasing motorization levels and a strong increase in trade and freight transportation (Box 11.4).

China became the world's largest automobile market in 2009, but the level of car ownership is still very low at 154 motor vehicles per 1000 people in 2016. That compares with the United States average of 797 vehicles per 1000 people. In India, transportation energy use is estimated to grow at 5.1%, the fastest rate in the world. Increase economic growth and an expanding population explain the rapid demand for passenger and freight transportation. Car registration in India grew threefold from 81 million in 2006 to 210 million in 2015 (Ministry of Statistics, Government of India, 2018). It is expected that the demand for personal car ownership will continue to increase at a rapid pace in the future.

In contrast, transportation sector energy use is estimated to decline in the OECD countries. The decline reflects the economic conditions of a mature market with slow economic growth and declining population, but also considerable improvement in energy efficiencies. Other key factors are high fuel costs partly due to the increase in taxation, government mandates on biofuels, and the switch to more fuel-efficient cars.

Africa's lack of infrastructure development and low income per capita appears to hold back the demand for energy transportation, which is forecast to grow at just 0.8% for the period between 2015 and 2040. In Central and South America, investment in road infrastructure is starting to pay dividends with an annual increase of 1.3% in energy

---

BOX 11.4 Global gas prices are converging since 2005

In the period 2005 to 2017, it is generally recognized that the global natural gas markets have become more integrated through increased LNG trade, increased market-related pricing, and gas hub development. It might be expected, therefore, that global gas prices would converge, as markets became more integrated. However, the conventional wisdom is that of global gas price divergence, based on a few regional gas price markers, such as Henry Hub, NBP [the UK's National Balancing Point, a virtual trading hub], and Japanese import prices.

However, analysis, using the price survey data, shows that global gas prices have been converging continuously since 2005, indicating further globalization of the gas markets. It is also concluded that there is more gas price convergence among countries (1) with market-related pricing, (2) which are connected with the global gas market through gas imports, (3) with oil-indexed gas prices, and (4) within Europe.

The trend of global gas price convergence is more distinct when we exclude the data of the effectively disconnected, North American countries. The pace of price convergence is even faster among countries with regulated gas prices than among countries with market-related prices. This is an indication that prices in countries with regulated prices are getting more aligned with global gas prices, probably through the elimination of subsidies and the increase of prices to more economical levels.

Digest of UK Energy Statistics (DUKES): Renewable Sources of Energy, n.d. DUKES chapter 6: statistics on energy from renewable sources. IGU Wholesale Gas Price Survey 2017 Edition. Available at: https://www.igu. org/sites/default/files/node-document-field_file/IGU_Wholesale%20Gas%20Price% 20Survey%202018%20Final.pdf. (Accessed 20th October 2018).

transportation estimated for the next 2 decades. Robust GDP growth, highly urbanized, and expanding young population have been crucial factors in the increase in transportation demand in the region.

According to the EIA, non-OECD Europe and Eurasia show an annual increase rate of 1.5% for energy transportation, higher economic growth, and an increase in disposable income is leading to higher demand for personal motor vehicles, particularly in former Soviet countries. In Russia, the largest economy in the region, transportation energy grows by 1.3% annual average from 2010 to 2040 despite population decline of 0.2%. The growth in energy transportation is driven by expanding car ownership and growth in freight transportation (IEA, International Energy Outlook, 2017).

The estimated growth in the transportation sector in developing countries is moderated by significant changes in consumption trends, such as increasing levels of energy efficiency and switch to hybrid and EV vehicles. The next revolution in energy consumption is likely to take place in the transport sector with the expansion of electric vehicles (EVs). The numbers are expected to grow from the current two million to 25 million in 2030 and 300 million in 2040 (IEA, World Energy Outlook, 2017). Efficiency improvements and innovative technologies not only reshape the transport sector but impact demand side of the energy business; several important developments are underway on the supply side, as well with car manufacturers focusing on developing electric or hybrid cars, some declaring that they will not be producing any combustion fuel cars by 2030. However, the switch to electric cars has a number of hurdles, including vehicle costs, charging infrastructure, power range, and battery issues. The overall cost of batteries are estimated to decline faster than initially expected with the price of electric cars following the same downward trend; to what extent electric cars become competitive versus conventional fuel cars will depend on commercial considerations, technological innovation, as well as government policies, particularly in respect to energy and carbon taxation.

According to the EIA, energy consumption from the building sector will increase by 32% during the forecast period from 2015 to 2040 — this accounts for 21% of the total world energy consumption by the end-user sector. Most of the growth will take place in non-OECD countries, which are experiencing substantial population growth, rapid economic development, and urbanization. Most of the increase in energy sources for both residential and commercial buildings will come from natural gas, which will continue to replace coal for heating and cooking.

11.14 The power sector

The power generation sector has been experiencing considerable changes driven by government policy and the pursuit of environmental agendas around the world. However, one of the most important factors in the power supply is reliability, and governments and regulators work together to ensure there is a legal and institutional framework to guarantee a secure and reliable supply of power with reduced environmental impact. Western countries have started to reduce emissions from the power sector by moving from coal power generation to natural gas and renewable energy. Global commitments toward climate change and de-carbonize energy is behind the drive toward clean power generation; wind power and solar power projects are becoming the fastest growing sectors in power generation. However, in terms of volume, natural gas is still the single leading generator by energy output.

According to the International Energy Agency, renewable energy, including hydro, was the third-largest source of power production in 2018, accounting for 26% of world electricity generation versus 74% from nonrenewables, mainly natural gas, coal, and nuclear energy. In spite of the fast pace of development of renewable energy on a global scale, the International Energy Agency (IEA) estimates that the total renewables share of global electricity generation, including hydro, is estimated to be 40% by 2040 with only 28% excluding hydro. This is because coal and natural gas will continue to be low-cost and reliable sources of power. However, the IEA outlook for renewables becomes more buoyant should the world adopt more aggressive action on climate change and may account for 37% of total electricity consumption by 2040 and representing around 50% by 2050. But even under such a hypothetical scenario, fossil fuels are still very much the dominant energy source in the next 2 decades, with gas accounting for a steady 20% and coal a 25% share by 2040 (IEA, WEO, 2018).

Most of the developed OECD countries are investing in renewable energy and gradually switching out of fossil fuels. However, in developing countries, the greatest problem is access to energy; lack of access to electricity continues to affect over one billion people, mostly from sub-Saharan Africa and Asia. For the population of these countries affected by no access to energy, renewables can offer a practical and affordable solution, particularly for those in rural areas away from the centralized grid. A significant expansion of off-grid solar PV systems has been taking place in developing countries facilitated by international multilateral and bilateral financing institutions.

11.15 The issues of rural electrification

The contrast between developed OECD countries and developing countries is quite sharp concerning access to energy, with the latter experiencing around 76% rate of access versus 99% in the former group of developed countries according to data from the World Bank (World Bank, 2008). Nearly one billion people worldwide have no access to electricity, and an additional two billion have inadequate access. The problem of lack of electrification is particularly acute in the regions of sub-Saharan Africa. According to the IEA, the electrification rate in this region is barely higher than a third, and rural areas show even lower levels of electrification. There is considerable evidence of a close connection between access to energy, particularly electricity and development. Chaurey et al. (2004) identified a link between rural poverty and lack of access to electricity. Other authors such as Khandker et al. (2009a,b) have stated that lack of energy access is an obstacle to economic development. While the relationship between energy access and development prospects appears to be widely accepted, the impact of electrification on the promotion of standards of life, health, and education in rural areas remains largely underresearched.

Rural areas, particularly in developing countries, are at a considerable disadvantage compared to urban areas. This is due to many factors, such as the low density of population, low-income level, which raises the issue of affordability, remote locations, and associated logistics and infrastructure obstacles, which challenges the feasibility of large power grid projects in these regions. A study by the World Bank (IEG, 2008) found that in spite of all the difficulties related to setting up rural electrification in rural areas, there is a positive impact on productive activities, health, and education within the community, as well as good returns on investment. However, it is perhaps fair to say rural electrification alone will not trigger development if not accompanied by investment in health, education, and promotion of agricultural activities on a commercial scale with the availability of water supply, secure land tenure, etc. Economic benefits connected with the use of digital and industrial technologies are also quite evident to promote living standards and economic opportunities opened by information technologies.

Another problem with lack or inadequate access to energy is connected with the use of highly polluting sources of energy used for heating and cooking. According to the World Bank, around 2.8 million people in the developing world rely on solid fuels such as wood and coal for heating and cooking. This has a detrimental effect on the health of households. In addition, it is estimated by the World Bank that an additional 200 million people use highly toxic kerosene for heating and cooking.

The benefits of electrification with respect to the health of the communities appear to have been well established. This is well illustrated in the Indian state Chhattisgarh, which has invested in solar to power around 900 health centers; this led to a significant improvement in health care and a decline in the number of deaths in the region.

The choice of energy source appears to be critical in terms of chances of successful implementation too. Major grid infrastructure projects are complex and costly, and most poor developing countries will find it difficult to fund these projects. Financing processes are key elements linked to sources of funding and how these funds are dispensed. The most successful processes appear to have incorporated recovery of costs; the case of Costa Rica, Thailand, and Tunisia illustrates this factor really well. On the other hand, in places where rural electrification is dependent on subsidies and donor programs, progress has been quite convoluted. The first step to evaluate the feasibility of electrification projects is to assess if conditions are in place to justify the investment, such as potential demand requirements. The choices of infrastructure projects are also dependent on resources available but also local conditions. In remote areas of difficult access, off-grid technology such as photovoltaic systems may be a more suitable and economical solution for rural electrification.

11.16 Off-grid electrification options

Mini-grid and stand-alone systems are types of off-grid options which appear to be quite appropriate for rural electrification, particularly in remote or difficult to access locations where grid transmission networks are not feasible investments. Mini-grids can serve small dwellings, either private or business customers supplying power with comparative quality to national grid networks. Mini-grids are often powered by clean energy, wind, solar, or biomass with back-ups in the form of batteries or conventional diesel. Similar to mini-grids, stand-alone systems are not linked to the national grid networks and supply individual homes or small businesses more often using renewable sources than conventional fuel sources such as diesel. These two options, in recent years, have become cost-competitive versus conventional combustion energy sources due to the falling costs of renewable technologies such as solar PV. The cost to produce power from solar PV, for example, dropped by 73% in the period between 2010 and 2017 (IRENA, 2017). At the same time, innovative solutions on models of financing and systems of implementation involving a wider range of stakeholders have created favorable conditions for the dissemination of these technologies. The decline in the cost of technology goes hand in hand with the surge of innovation in financing solutions for power consumers. For example, pay-as-you go associated with power supply in East Africa, and microfinance solutions in Bangladesh have achieved remarkable implementation success. It is estimated by the IEA that 60% of new electricity access will derive from renewable sources by 2030, and half of it will be mini-grids or stand-alone systems; most of this demand derives from the developing countries of Asia and Africa. A combination of private sector investment and the support of government have enabled successful mini-grid investment in countries of sub-Saharan Africa and Southeast Asia. However, there is the need to attract commercial debt finance as the majority of mini-grid renewable projects are still anchored on grants and non-commercial loans, such as community funding and donation funds. The support of the government to promote off-grid systems is also crucial in the form of energy policy and regulatory measures such as fiscal incentives and reform of fossil fuel subsidies. The government can also play a pivotal role in supporting the financial viability of off-grid systems, as well as looking out for the interest of potential customers to ensure affordable power tariffs.

Main socioeconomic benefits from micro-grid or stand-alone systems using renewable energy:

- Cost competitive
- Job creation
- Development of local skills
- Unlocking local socioeconomic development potential
- Enhancing health and well-being by reduction of polluting sources from heating and cooking within the homes

As most of the benefits are self-explanatory, it is worth noting that job creation and development of local skills are crucial aspects to guarantee successful implementation, as local knowledge of the use of renewable technology enables feasible long-term operations. These potential employment opportunities can be promoted along the value chain. There is still significant scope for improvement in renewable technologies, particularly on storage and digital technology, with the potential to reduce the costs of renewable energy further in the future.

11.17 Levelized Costs of electricity

It is common practice to use the levelized cost of electricity (LCOE) to calculate the production cost of the various sources of energy for power generation on an equal basis. This tool measures the competitive cost of the electricity-generating system during its lifetime, including operations, maintenance, cost of raw feedstocks, and investment cost. This can be defined as the net present value of all costs over the lifetime of the assets divided by the total electrical energy output of the specific asset (Fig. 11.13).

The levelized cost of electricity (LCOE):

$$LCOE = \frac{\text{sum of costs over lifetime}}{\text{sum of electrical energy produced over lifetime}} = \frac{\sum_{t=1}^{n} \frac{I_t + M_t + F_t}{(1+r)^t}}{\sum_{t=1}^{n} \frac{E_t}{(1+r)^t}}$$

FIGURE 11.13 Levelized cost of electricity (LCOE).

The levelized cost of electricity (LCOE):

A recent study by the Financial Advisory firm Lazard on the levelized cost of electricity in the US markets illustrates the significant drop in the cost of renewable energy for US power generation but also demonstrates the competitive level of energy alternatives versus conventional energy in the country of analysis. (Levelized Cost of Energy Analysis, Lazard, 2017). Other studies conducted across developed countries arrive at the same conclusions. A German LCOE study was carried out by Fraunhofer Institute and points out to similar findings.

LCOE is a useful tool for comparing competing sources of power generation; however, it should not be the only source of analysis. Comparing LCOE across technologies can be misleading if used as a single method of evaluating the economic competitiveness of competing power generation; factors such as utilization rates, capacity values, existing resources mix must be considered as these factors vary across regions. In addition, financing factors such as interest paid, leasing versus ownership, maintenance costs, replacement parts, are all factors that potentially impact the LCOE of a project.

Most analysts have agreed on the dominant role of oil and gas lasting well into the 2040s. Most also agree that natural gas will displace coal as the second-largest fuel after oil. However, energy efficiencies across transport, industrial, and power sectors will significantly reduce growth rates for fossil fuels from all regions of the world. Renewable energy and energy efficiencies are key elements for the successful implementation of the low-carbon energy transition. The declining costs of decentralized renewables and increased access to energy efficiency tools are changing the energy landscape and opening up opportunities for further diversification of energy resources enhancing the security of supply. The government can play a significant role in promoting energy policies that spur economic growth without compromising climate change goals (Box 11.5).

11.18 Conclusion

Global energy markets are being shaped by evolving energy transition challenges where patterns of demand are changing, and the mix of energy supply is facing the dual goal of meeting required energy needs while also mitigating carbon emissions. Energy outlooks from a range of energy forecasts indicate energy from fossil fuels is still going to be dominant for the foreseeable future, but the pace of transition toward cleaner and

---

BOX 11.5 Case study: Germany

In 2013, The Fraunhofer Institute for Solar Energy Systems ISE carried out an assessment of the levelized generation costs for newly built power plants in the German electricity sector. The study compared the cost of producing power from a range of power generation systems, including Solar (PV), onshore and offshore wind, coal, biogas, and natural gas. The data has been updated to keep up with industry developments. The latest data published was in 2018 and demonstrates that onshore wind projects and PV are competitive versus conventional energy generation. The study shows that PV systems have an LCOE between 3.71 and 11.54 €Cents/kWh, while the LCOE of onshore wind turbines ranges between 3.99 and 8.23 €Cents/kWh. LCOE of offshore wind turbines from 7.49 to 13.79 €Cents/kWh is significantly higher than onshore wind turbines. Except for offshore wind projects, PV systems and a significant number of wind projects are cheaper than conventional power generation such as coal and CCGT natural gas plants. Therefore, those compare favorably with the conventional power plants. The LCOE for the cheapest conventional brown coal lies between 4.59 and 7.98 €Cents/kWh.

The combination of increased scale and technological innovation has enabled the achievement of grid parity in Germany for renewable energy. The high cost of offshore wind represents a small fraction (3%) of total wind capacity; the high costs of offshore represent high commercial, environmental, and regulatory barriers to the installation of offshore wind turbines. In Germany, solar represents the fastest growing renewable energy sector. The development of solar power is the result of rapid improvements in solar photovoltaic technology, notably, advances in crystalline-silicon PV that have produced a dramatic decline in the cost of solar electricity.

---

more environmentally friendly energy is still uncertain. In spite of some antagonistic rhetoric between "renewable energy" and "fossil fuel" camps, it is becoming increasingly clear that both sectors must make a crucial contribution to fulfill future energy demand. A good dose of pragmatism is required to recognize the key role of fossil fuels during the transition period; investment in oil and gas continues to be an attractive proposition, but it is also a necessity to meet the energy needs of future global energy markets.

Most analysts have agreed on the dominant role of oil and gas lasting well into the 2040s. Most also agree that natural gas will displace coal as the second-largest fuel after oil. However, energy efficiencies across transport, industrial and power sectors will significantly reduce growth rates for fossil fuels from all regions of the world. Renewable energy and energy efficiencies are key elements for the successful implementation of the low-carbon energy transition. The declining costs of decentralized renewables and increased access to energy efficiency tools are changing the energy landscape and opening up opportunities for further diversification of energy resources, enhancing the security of supply. The government can play a significant role in promoting energy policies that spur economic growth without compromising climate change goals.

The energy sector and energy markets are being swept by profound changes affecting the way we produce, consume, and manage our energy resources. These changes will make a significant impact on the economies of oil-producing countries and the portfolio of oil and gas companies. This period of energy transition marks the beginning of a new era where fossil fuel is being challenged as the dominant source of energy, and new energy alternatives are increasingly gaining market share within the energy mix. Global commitment toward environmental targets and the reduction of emissions will require substantial changes in the energy landscape with wide implication worldwide for both consumer and oil-producing countries. International oil and gas companies are beginning to realize the massive challenges ahead and are starting to prepare by integrating some low-carbon assets into their portfolio. However, to date, such investment by oil and gas companies are still very timid compared to the overall investment budget for fossil fuels. The nature and time scale of the energy transition period is also uncertain and brings critical dilemmas for companies, as well as oil-producing countries, regarding how fast the transition period will take with implications for strategic decision-making such as potential diversification of energy investment, and what technologies will be deployed.

Nevertheless, it is becoming clear the trajectory toward clean energy is irreversible with global environmental commitments and domestic environmental policies increasingly pursuing clean energy, particularly in developed countries. Developing countries are also starting to connect with international efforts led by the United Nations toward investment in clean energy as the link between environmental objectives and economic benefits, has become better understood and relatively well established.

The arguments in favor of investment in clean energy have been gaining ground as renewable energy is becoming cost-competitive versus conventional energy making good commercial sense, as well as being a social and environmental sound investment.

In spite of the bullish forecast concerning the growth of renewable energy, we should not be forecasting the demise of the oil sector just yet. In reality, oil production will continue to rise to meet demand, rising from 100 million barrels per day to 125 million by 2040, according to the EIA (International Energy Outlook, 2017). Oil demand growth will, however, be more moderate, with energy efficiencies being one of the factors slowing demand growth followed by growth in electric vehicles in the next decades.

During the transition period, natural gas is becoming more relevant as the cleanest source of energy among the fossil fuel sector. Expansion of LNG production capacity in Australia, the USA, and Africa does ensure natural gas becoming a globally tradable commodity. Natural gas will continue to be the single leading power generation by energy output for the next decade in spite of significant inroads from renewable energy.

The power generation sector will continue to experience considerable changes driven by government policy and the pursuit of environmental agendas around the world. However, reliability and security of supply will be the driving considerations for governments and regulators in the power sector. Western countries have managed to reduce emissions from the power sector by moving from coal power generation to natural gas and renewable energy. Global commitments toward climate change and decarbonization of energy are behind the drive toward clean power generation. Energy efficiencies in transport, industrial, and the building sector are estimated to reduce growth rates for fossil fuel from all regions of the world. The government will play a key role in promoting energy policies that reconcile climate change goals with socio and economic development objectives.

In the case of rural locations in developing countries, the choice of technology depends, as mentioned earlier, on resources available but also specific conditions such as population density and dispersion. Stand-alone diesel or renewable energy off-grid options have been proved to achieve low LCOEs, and therefore, to be more cost-competitive in the low-density population with high dispersion rate. By contrast, in locations with a higher density population, mini off-grid options appear to achieve lower LCOEs both for diesel-powered and solar-powered systems but with renewable energy having the advantage to be more sustainable long-term and more environmentally friendly.

References

Australian Department of industry, Innovation and Science, n.d. Australia LNG Export Capacity. Available at: https://publications.industry.gov.au/publications/resourcesandenergyquarterlyjune2018/index.html. (Accessed 30th August 2018).

BP Statistical Review of World Energy, 2018. Renewable Energy by Region. Available at: https://www.bp.com/en/global/corporate/energy-economics/statistical-review-of-world-energy/downloads.html (Accessed 10th July 2018).

Chaurey, A., Ranganathan, M., Mohanty, P., 2004. Electricity access for geographically disadvantaged rural communities–technology and policy insights. Energy Pol. 32 (15), 1693–1705.

Digest of UK Energy Statistics (DUKES): Renewable Sources of Energy, n.d. DUKES chapter 6: statistics on energy from renewable sources. IGU Wholesale Gas Price Survey 2017 Edition. Available at: https://www.igu.org/sites/default/files/node-document-field_file/IGU_Wholesale%20Gas%20Price%20Survey%202018%20Final.pdf. (Accessed 20th October 2018).

EIA Database, n.d. Brent Spot Price. Available at: https://www.eia.gov/dnav/pet/hist/LeafHandler.ashx?n=PET&s=RBRTE&f=M. (Accessed 30th August 2018).

EIA, Database, 2018. WTI Crude Prices versus Henry Hub Natural Gas Prices. Available at: https://www.eia.gov/outlooks/aeo/data/browser/ (Accessed 17th July 2018).

EIA, International Energy Outlook, 2017. Total Primary Energy Consumption. Available at. https://www.eia.gov/outlooks/ieo/pdf/0484(2017).pdf (Accessed 17th July 2018).

European Environmental Agency, 2017. EU 28 Energy Intensity Per Unit of GDP. Available at: https://www.eea.europa.eu/data-and-maps/indicators/total-primary-energy-intensity-3/assessment-1 (Accessed 10th August 2018).

IEA, Energy Investment, 2017. World Energy Investment 2017. Available at: https://www.iea.org/reports/world-energy-outlook-2017#global-shifts-in-the-energy-system (Accessed 8th July 2018).

IEG, 2008. The Welfare Impact of Rural Electrification: A Reassesment of the Costs and Benefits. The Independent Evaluation Group, The World Bank Group, Washington, D.C. http://documents1.worldbank.org/curated/en/317791468156262106/pdf/454000PUB0978011PUBLIC10Mar06102008.pdf.

International Energy Outlook, 2017. (EIA). Available at: https://www.eia.gov/outlooks/ieo/pdf/0484(2017).pdf.

International Gas Union (IGU), 2019. Available at: https://www.igu.org/resources/lng-2019-report/.

Khandker, S.R., Barnes, D.F., Samad, H.A., 2009a. Welfare impacts of rural electrification: a case study from Bangladesh. In: Policy Research Working Paper Series 4859. The World Bank.

Khandker, S.R., Barnes, D.F., Samad, H.A., Minh, N.H., 2009b. Welfare impacts of rural electrification: evidence from Vietnam. In: Policy Research Working Paper Series 5057. The World Bank.

LNG World News, 2016. Available at: https://www.offshore-energy.biz/lngworldnews/.

Ministry of Statistics, Government of India, 2018. Available at: http://mospi.nic.in/statistical-year-book-india/2018/189.

Renewable Energy Statistics, IRENA, 2017. Available at: https://www.irena.org/publications/2017/Jul/Renewable-Energy-Statistics-2017.

Shell.com, 2019. Available at: https://www.shell.com/promos/download-the-full-lng-2019/_jcr_content.stream/1550845695544/58c13c03d98ad9a29a6a27b5beab32d58b4ced34/lng-outlook-factsheet.pdf.

The Fraunhofer Institute for Solar Energy Systems ISE, n.d. Levelised Cost of Electricity. Available at: https://www.ise.fraunhofer.de/content/dam/ise/en/documents/publications/studies/EN2018_Fraunhofer-ISE_LCOE_Renewable_Energy_Technologies.pdf. (Accessed 30th August 2018).

United Nations Population, 2018. United Nations, Department of Economic and Social Affairs, Population Division (2018). World Population Prospects: The 2018 Revision. New York: United Nations. Available at: https://population.un.org/wup/Download/.

Vygon consulting, n.d. LNG Production Capacity 2015 and 2020. Available at: https://vygon.consulting/upload/iblock/3cd/vygon_consulting_us_lng_2017_en.pdf. (Accessed 1st October 2018).

12

Energy technologies and energy storage systems for sustainable development

In today's world, there is a continuous global need for more energy, which at the same time, has to be cleaner than the energy produced from the traditional power-generation technologies. This need has facilitated the increasing penetration of distributed generation (DG) technologies and primarily renewable energy sources (RES). New discoveries and innovations have driven the development of the energy system since the 18th Century. For many rural communities, there is no immediate prospect of being connected to the central electricity grid, and other commercial energy sources are often too expensive for poor people. Global energy consumption produces a large share of the carbon dioxide emitted to the atmospheric emissions that affect our climate and contribute to global warming; meanwhile, global energy demand continues to rise (Byrne et al., 2006). All energy sources have some type of climate or environmental impact during their life cycle. In the short term, emissions from existing power plants must be minimized. Alternative energy, as it is currently conceived, is that which is produced or recovered without the undesirable consequences inherent in fossil fuel use, particularly high carbon dioxide emissions, a key factor in global warming. While public debate about alternative energy has focused primarily on wind and solar power, far more dramatic and far-reaching solutions already exist or are in exploratory stages. The public knows little about these alternative energy sources, which energy cartels and governments have intentionally and sometimes brutally suppressed. These newer solutions, such as energy extracted from water, energy captured from background energy of the cosmos (called zero-point energy, the vacuum, or etheric energy), energy generated by magnetic fields, vortex energy, and cold fusion, often defy the concepts of familiar physics. Because poor people in rural areas lack access to electricity and modern fuels, they rely primarily on human and animal power for mechanical tasks, such as agricultural activities and transport, and on the direct combustion of biomass (wood, crop residues, dung) for activities that require heat or lighting. These energy sources may offer a far more complete and long-term solution to our energy problems; all are totally available and environmentally friendly, producing no pollution. This chapter will be reviewing some of the main alternative sources of energy being proposed for the survival of our planet.

12.1 Rural areas and power supply

Modern energy forms are an economic good, capable of improving the living standards of billions of people, particularly in developing countries, who lack access to service or whose consumption levels are far below those of people in industrialized countries. By using decentralized energy systems, the high costs associated with transmission and distribution through the national grids can be avoided. The possibility of adopting renewable energy technology (RET) is particularly important in light of the limited success of conventional rural electrification programs that have been designed to meet energy needs in developing countries.

The absence of commercially supplied energy in society, especially electricity, tends to accentuate the existence of social asymmetry in living conditions. This can take the form of increased poverty, lack of opportunity for development, migratory flow to large cities, and a society's disbelief regarding its own future. There is a general belief that, with the arrival of electricity, such societies might acquire a higher degree of economic sustainability and a better quality of life (Pereira et al., 2010). It is estimated that one-third of the world's population has no access to electric energy; half of this population lives on the African continent, many in rural locations. On the other hand, one of the paths toward economic sustainability refers to the availability of access to regular electric energy. Such access is a key element for the economic development of the rural environment and for the reduction of poverty. However, expanded access to electric energy has shown slow progress worldwide, especially due to the high costs associated with extending grids and developing decentralized systems that offer power (Zhou et al., 2008).

As part of these efforts to provide electricity off the grid, in many cases by the adoption of renewable technology systems—the most common being diesel-engine generator set, small-scale hydropower, photovoltaic (PV), wind, and small-scale bio-power producing gas—have been installed across the developing world in Latin America, the Caribbean, Africa, and Asia in the past 4 decades. However, only a minority of the rural population without electricity has access to some form of modern renewable energy services; that is, more than 2.4 billion people worldwide continue to rely fully on traditional fuels to cook (Howells et al., 2005).

12.2 The energy problems of modern societies

We all know that a secure energy supply is vital to modern society. Occasional power cuts are a nuisance, and sometimes rather more than that for rural households when power lines are down in winter. In the UK, plans to make deep cuts in emissions of greenhouse gases, now entrenched via the Climate Change Act of 2008, put the country's already vulnerable electricity supply under considerably greater pressure. At the same time, the existing means of energy production face new problems. International treaties aim to limit the levels of pollution, and global warming prompts action to reduce the output of carbon dioxide, and several countries have decided to decommission old nuclear power plants and not build new ones. In addition, the unprecedented global increase in energy demand has meant that the price of conventional energy sources has risen dramatically and that the dependence of national economies on a continuous and undistorted supply of such sources has become critical.

Research and development are still of paramount importance within the energy sector, particularly in terms of managing some of the major challenges facing mankind today, such as greenhouse gas emissions and global warming. New problems lead to new approaches, new discoveries to new opportunities, and hard work to continued improvements. Energy has both positive and negative impacts on societies. Access to abundant, affordable, secure, safe, and clean energy is beneficial for humans. But energy extraction, transportation, and use can have negative consequences on the health, environment, and economics of a society. Moreover, relying on imported energy can create vulnerabilities to a nation's security (Popescu, 2015).

The impacts of energy decisions are not equal for all people. Poor or marginalized societies are more likely to suffer the negative consequences of energy decisions because they have a reduced capacity for adaptation, and they may lack negotiating power compared to wealthier societies. Therefore, vulnerable populations can benefit greatly from improvements in energy accessibility, safety, or affordability. Access to energy is critical to sustainable development, to building stronger communities with hospitals and schools. It supports business and industry so that they can deliver sustainable employment and economic growth. Ensuring access to electricity and supporting economic growth is, in turn, essential to support human development. Without this, modern society cannot develop or function effectively. Energy should generate a consistent stream of power to meet basic human needs, maintain and improve social functioning, and advance living standards. It should also fulfill these functions as sustainably as possible, that is to say, the power generated by energy use should be much greater than the resulting waste and pollution. All sustainable energy must be modern, although not all forms of modern energy are sustainable.

Coal is perhaps the most important case in point. Historically, coal has been indispensable to industrialization and the advancement of human well-being. If more of the world's people enjoy previously unimaginable living standards today, it is in large part because of coal. Offsetting its many virtues—for instance, abundance, wide distribution, and ease of use—is a long list of serious problems. In an age of population growth and environmental decline, this list is still growing. Today, coal still provides about 40% of the world's electricity and nearly the same fraction of global carbon emissions. Coal is also inefficient, with a low mass-to-energy ratio, and creates enormous pollution. Thus, coal is neither sustainable at the global scale because of its contribution to anthropogenic climate change, nor at the local scale because it is a threat to public health and ecological conditions (in addition to the polluting by-products of combustion, the process of coal mining creates myriad environmental problems). These are challenges that require a pragmatic, multi-faceted approach. Solutions need to be found on a global scale, where Governments and agencies must work together.

International climate change agreements are the most visible fruits of such efforts. In terms of policies, the transfer of clean energy technologies to developing countries is an important example. Indeed, international climate change agreements—such as the clean development mechanism (CDM)—explicitly provide for such transfers. This is not enough, however, as solutions must also be developed locally. Development should be sensitive to local conditions and identify unintended consequences of energy policies. The heedless pursuit of biofuels at the global and regional levels may result in unintended yet severe environmental degradation. The countless acres of land deforested for palm oil undermine local well-being and provide a stark reminder of the complexity of the energy problems that we face (Box 12.1).

BOX 12.1 Nikola Tesla: electrical genius

Nikola Tesla was a Serbian-American inventor, electrical engineer, mechanical engineer, physicist, and futurist best known for his contributions to the design of modern alternating current (AC) electricity supply system. He was born in 1856 in what Croatia is now and attended the University of Graz in Austria in 1878 and 2 years later, the Charles-Ferdinand branch of the University of Prague in Bohemia, where, in both schools, he took courses such as integral calculus, geometry, analytical chemistry, machine construction, botany, wave theory, optics, philosophy, French, and English. Tesla went to the United States in 1884 and briefly worked with Thomas Edison before the two parted ways. Tesla filed more than 300 patents during his 86 years of life, and his inventions helped pave the way to alternating current (AC), electric motors, radios, fluorescent lights, lasers, and remote controls, among many other things. Without Tesla, we might not be able to affordably power our home, let alone having access to the Internet. Tesla was one of our greatest visionaries, very far in advance of his time. He remains as an inspiration to many in Silicon Valley today. A tremendous resurgence of interest in him has occurred because of the Internet, a global communication system very much in accord with his inventions and predictions.

BOX 12.2 Rural energy policies

Evidence suggests that people are willing to spend a significant portion of their incomes on high-quality energy that improves their quality of life and enables them to be more productive. Governments have an important role to play in creating conditions that provide consumers with more energy choices and encourage innovation and investment in new technologies. Prices should be liberalized to reflect costs, and regulatory policies need to encourage competition and level the playing field for all types of energy markets, whether they are served by public utilities, private firms, or community enterprises. For example, "off-grid" power companies and cooperatives are often totally excluded by electricity regulations from serving people, and policies that artificially hold down prices sometimes provide little incentive for such local initiatives to get started.

12.3 Renewable energy as a way out of the energy crises

Renewable technologies are considered as clean sources of energy, and optimal use of these resources minimize environmental impacts, produce minimum secondary wastes and are sustainable based on current and future economic and social societal needs (Divya and Jibin, 2014). Renewable energy resources will play a significant role in the world's future. Conventional energy sources based on oil, coal, and natural gas have proven to be highly effective drivers of economic progress, but at the same time damaging to the environment and to human health (Panwar et al., 2011). Furthermore, they tend to be cyclical in nature due to the effects of oligopoly in production and distribution. These traditional fossil fuel-based energy sources are facing increasing pressure on a host of environmental fronts, with perhaps the most serious challenge confronting the future use of coal being the Kyoto Protocol greenhouse gas (GHG) reduction targets (Galvez et al., 2010). It is now clear that any effort to maintain atmospheric levels of CO_2 below even 550 ppm cannot be based fundamentally on oil- and coal-powered global economy, barring radical carbon sequestration efforts. The potential of renewable energy sources is enormous as they can, in principle, meet the world's energy demand many times over.

Renewable energy sources such as biomass, wind, solar, hydropower, and geothermal can provide sustainable energy services based on the use of routinely available indigenous resources. A transition to renewables-based energy systems is looking increasingly likely as the costs of solar and wind power systems have dropped substantially in the past 30 years, and continue to decline, while the price of oil and gas continue to fluctuate. In fact, fossil fuel and renewable energy prices, social and environmental costs are heading in opposite directions (IRENA, 2018). Furthermore, the economic and policy mechanisms needed to support the widespread dissemination and sustainable markets for renewable energy systems have also rapidly evolved. It is becoming clear that future growth in the energy sector is primarily in the new regime of renewable, and to some extent, natural gas-based systems and not in conventional oil and coal sources. Financial markets are awakening to the future growth potential of renewable and other new energy technologies, and this is a likely harbinger of the economic reality of truly competitive renewable energy systems (Hadjipaschalis et al., 2009). Renewable energy system development will make it possible to resolve the presently most crucial tasks like improving energy supply reliability and organic fuel economy; solving problems of local energy and water supply; increasing the standard of living and level of employment of the local population; ensuring sustainable development of the remote regions in the desert and mountain zones; implementation of the obligations of the countries with regard to fulfilling the international agreements relating to environmental protection.

Climate Change is one of the primary concerns for humanity in the 21st Century (Willcox, 2012). It may affect health through a range of pathways, for example, as a result of increased frequency and intensity of heatwaves, reduction in cold-related deaths, increased floods, and droughts, changes in the distribution of vector-borne diseases and effects on the risk of disasters and malnutrition. The overall balance of effects on health is likely to be negative, and populations in low-income countries are likely to be particularly vulnerable to the adverse effects. Development and implementations of a renewable energy project in rural areas can create job opportunities, and thus, minimizing migration toward urban areas. Harvesting renewable energy in a decentralized manner is one of the options to meet the rural and small-scale energy needs in a reliable, affordable, and environmentally sustainable way.

12.4 Alternative energy sources: development & prospects

With humanity facing the dismaying prospects of global ecological collapse and geopolitical chaos, there is an urgent need for clear solutions-based guidance that penetrates our dulled consciousness and pulls us back from the precipice. In the future, civilization will be forced to research and develop alternative energy sources. Our current rate of fossil fuel usage will lead to an energy crisis this century. The hope is that millions will heed this call for action without delay and lead the transformation so desperately needed on our imperiled planet. The following are some possible alternative energy technologies and sources.

12.4.1   Solar energy collection using a spherical sun power generator

Few years ago, futurist Ray Kurzweil predicted that within 20 years, solar power technology would move forward to the point where it would have the capabilities to supply all of the world's energy needs. This forecast was no exaggeration of our abilities considering that the amount of energy Earth receives in just 1 h would be enough to power humanity over the course of a year, making solar the only renewable energy that can keep up with global demands (Cass, 2009). This technology was designed by the German Architect Andre Broessel who believes he has a solution that can capture more energy from the sun event during the night hours and in low-light regions. This technology will combine spherical geometry principles with a dual-axis tracking system, allowing twice the yield of a conventional solar panel in a much smaller surface area. The futuristic design is fully rotational and is suitable for inclined surfaces, walls of buildings, and anywhere with access to the sky. This technology can even be used as an electric car charging station. Currently, photovoltaic (PV) based solar panels have been in the commercial-use phase for some time and are available in hardware stores. Some advantages of spherical shape are larger exposure to the sun's rays, which can result in large amount of electricity be produced, leading to maximum use of power plants. Due to the continuous periodic rotation of solar power plant gravitation inside, this tool can be maintained much effectively.

12.4.2 Space-based solar power (SBSP)

This technology resource is a concept in which satellites in Earth's orbit are able to capture solar power in outer space, convert it to electricity and then distribute it back to Earth in the form of either microwaves or lasers. The are several advantages of collecting radiation from the Sun in space, and these include a higher collection rate and a longer collection period due to the lack of diffusing atmosphere, as well the possibility of placing a solar collector in an orbiting location with no night, thus able to continuously collect solar power. Averaged over day/night and a full year, solar radiation delivers to Earth's surface a continuous flux between 100 and 300 W/m^2 (Seboldt, 2004), depending on the region. It appears that larger values are obtained closer to the equator. The transformation into useful forms of power is typically performed via solar photovoltaic or solar thermal devices, where conversion efficiencies to the electricity of 15%−20% are achieved or can be assumed for the near future (The Renewable Energy Hub, 2018). SBSP is considered as a form of sustainable or green energy, renewable energy and is often considered among climate engineering proposals.

12.4.3 Algal biofuel

Algal biofuel, also known as Algae fuel or algal oil that uses algae as its source of energy-rich oils, is an alternative to liquid fossil fuels. The term "algae" refers to a great diversity of organisms from microscopic cyanobacteria to giant kelp, which converts sunlight into energy using photosynthesis, like plants (Vidyasagar, 2016). Algae, particularly green unicellular microalgae have been proposed for a long time as a potential renewable fuel source. Algae fuels are an alternative to commonly known biofuel sources such as corn and sugarcane (Pittman et al., 2011). Microalgae have the potential to generate significant quantities of biomass and oil suitable for conversion to biodiesel. Microalgae have been estimated to have higher biomass productivity than plant crops in terms of land area required for cultivation, are predicted to have lower costs per yield, and have the potential to reduce greenhouse gas (GHG) emissions through the replacement of fossil fuels. Many species of microalgae are able to effectively grow in wastewater conditions through their ability to utilize abundant organic carbon and inorganic Nitrogen and Phosphorous in the wastewater (Abedel-Raouf et al., 2012). Although the application of microalgae in the wastewater industry is still fairly limited, algae are used throughout the world for wastewater treatment, albeit on a relatively minor scale. Large-scale commercial production of algae, however, is potentially more costly than traditional crop production. Algae cultivation requires an abundance of water and nutrients, such as carbon, nitrogen, and phosphorus. The application of external nutrient sources results in direct competition for fertilizers with food growers. In order to make microalgae-based biofuel cost-effective, wastewater and flue gas can be employed as cheap nutrients sources. Such a method would reduce the reliance on chemicals providing the nutrients, making it also ecologically favorable.

12.4.4   Tidal power

Oceans cover over two-thirds of our planet's surface. Advancements in engineering through human civilization have sought to take advantage of this huge abundance of water. Tidal power or tidal energy is a form of hydropower that converts the energy obtained from tides into useful forms of power, mainly electricity. Leaps in engineering have far outstripped these technologies; however, the idea of extracting energy from water has remained. The methods of doing so can be broken down into three main categories: wave energy, tidal stream, and tidal range.

Tidal range power is created using a head difference between two bodies of water (Waters and Aggidis, 2016). For creating this difference, a wall is used to separate the two areas, and as the tide flows in or out, the wall blocks the flow of the tide and creates a head difference. When the head difference has reached an optimum level, the water passes through the barrage and creates energy due to the turbines placed within the holes in the wall. Among sources of renewable energy, tidal energy has traditionally suffered from relatively high cost and limited availability of sites with sufficiently high tidal ranges or flow velocities, thus constricting its total availability. However, many recent technological developments and improvements, both in design (e.g., dynamic tidal power, tidal lagoons) and turbine technology (e.g., new axial turbines, cross-flow turbines), indicate that the total availability of tidal power may be much higher than previously assumed and that economic and environmental costs may be brought down to competitive levels.

Historically, tide mills have been used both in Europe and on the Atlantic coast of North America (Greaves and Iglesias, 2018). Tidal Energy has an expensive initial cost, which may be one of the reasons that tidal energy is not a popular source of renewable energy. It is important to realize that the method for generating electricity from tidal energy is a relatively new technology. It is projected that tidal power will be commercially profitable within 2020 with better technology and larger scales (Maehlum, 2013).

12.4.5   Nuclear waste

Nuclear waste is the material formed from nuclear fuel after it is used in a reactor. From the outside, it looks exactly like the fuel that was loaded into the reactor. But since nuclear reactions have occurred, the contents are not quite the same. Nuclear waste epitomizes the double-edged sword of modern technology. It is a toxic and radioactive byproduct of nuclear medicine, nuclear weapons manufacturing, and nuclear power plants. Radioactive waste is hazardous to all forms of life and the environment and is regulated by government agencies in order to protect human health and the environment. Depending on the waste's source, the radioactivity can last from a few hours to hundreds of thousands of years (Baisden et al., 2006). If disposed of improperly, radioactive waste can devastate the environment, ruining the air, water, and soil quality. What is more, these materials can have long-term negative effects on human health and can be fatal.

Nuclear power is characterized by the very large amount of energy produced from a very small amount of fuel, and the amount of waste produced during this process is also relatively small. However, much of the waste produced is radioactive, and therefore, must be carefully managed as a hazardous material. Radioactivity naturally decays over time, so radioactive waste must be isolated and confined in appropriate disposal facilities for a sufficient period until it no longer poses a threat. The time for storing radioactive waste depends on the type of waste and radioactive isotopes (Ellis, 2017). Current approaches to managing radioactive waste have been segregation and storage for short-lived waste, near-surface disposal for low and some intermediate-level waste, and deep burial or partitioning/transmutation for the high-level waste.

12.4.6   Solar windows

Solar windows are windows that function as solar panels to harvest the sun's energy and convert it into electricity. They are created by applying a photovoltaic film onto window glass (Chow et al., 2010). Some manufacturers create independent solar windows by sandwiching existing solar panel cells between two layers of glass. Although the solar cells are visible, the human eye skips over the cells when looking out of the window. Photovoltaic films allow building owners to modify their existing windows, installing a film on the inside surface of the glass (Alameh et al., 2014). The film is made from mostly organic materials, such as carbon, nitrogen, hydrogen, and oxygen, that are applied in liquid form to a stabilizing layer. Solar windows would hypothetically be able to replace standard glass window-panes, while traditional solar panels are an addition to a previously installed roof. As a result, this type of solar technology is often referred to as "building-integrated photovoltaics (BIPV)."

12.4.7   Human power

About a century and a half ago, products that relied on human energy such as the bicycle, pedal-powered lathe, or sewing machine could be found in most households. But as electro-mechanical motors developed, reliance on human-powered products gradually diminished. Today, human power is not appropriately recognized for its potential as an alternative solution to our growing energy needs. The human body contains enormous quantities of energy. In fact, the average adult has as much energy stored in fat as a one-ton battery. That energy fuels our everyday activities, but what if those actions could, in turn, run the electronic devices we rely on? Movement produces kinetic energy, which can be converted into power (McClelland, 2015). In the past, devices that turned human kinetic energy into electricity, such as hand-cranked radios, computers, and flashlights, involved a person's full participation. But a growing field is tapping into our energy without our even noticing it. Unlike solar and wind energy, human power is always available, no matter the season or time of day. Unlike fossil fuels, human power can be a clean energy source, and its potential increases as the human population grows. Human-powered products also have the potential to encourage us to become more physically active. Using human-powered products as a countermeasure to our increasingly sedentary lifestyles could create a credible new perspective toward exercise as an alternative energy source. In some respects, human-power can be seen as the cleanest renewable energy source available, with great potential for helping people stay healthy and have fun.

12.4.8 Flying wind farms

Wind speeds increase with height, and for several years, engineers have been looking at ways to harvest wind power and convert it into cheap and renewable energy Liserre et al. (2010). Flying wind farms are air turbines similar to the conventional tower-mounted turbines able to handle the wind with high speed with additional devices fitted on it. They will automatically detect and adjust their height to use the best wind speed (Kalogirou, 2005). Flying wind farms have advantages over their land-based counterparts because of factors such as contours of the land and daily heating and cooling patterns. No such impediments occur in the jet stream, where air moves near constantly and at several times the speed that it does at 100 feet off the ground, allowing much more energy to be captured from each square meter of wind. Flying wind farms are even more advantageous as it has ad-hoc generation; devices with reasonably simple tether-system that do not have to be permanently installed in one place (Bolonkin, 2004). The principle is similar to conventional tower-mounted turbines in its working, but here the rotor and generator will be floating in the air just like a hot air balloon. The most important disadvantage of conventional types of wind farms is that there can be times when there is not enough wind, whereas, at higher altitudes, wind conditions are much more consistent.

12.4.9 Nuclear fusion

Nuclear fusion can be defined as a nuclear reaction, in which lighter nuclei are combined together to form heavier product nuclei with the release of an enormous amount of energy. The vast energy potential of nuclear fusion was first exploited in thermonuclear weapons or hydrogen bombs, which were developed in the decade immediately following World War II (Conn, 2018). Meanwhile, the potential peaceful applications of nuclear fusion, especially in view of the essentially limitless supply of fusion fuel on Earth, have encouraged an immense effort to harness this process for the production of power.

12.5 Energy storage systems

Today's world, and particularly those developing countries, relies heavily on fossil fuels. Most of the time the fossil fuels are consumed for heat and electricity. The growing world population and increasing standard of lifestyle have led to a rapidly increasing demand for energy since 1950 and are projected to peak in 2035 (Li, 2014). Moreover, the nonrenewable nature of fossil fuels such as coal, oil, and natural gas at the humankind timescale has prompted governments in many countries to think about energy security. Energy in any form is an essential commodity globally. It is the most common consumer good and has continued to be a key element to worldwide development.

Fossil fuels will oneday be inevitably used up, although this may not occur in the next two generations due to the relatively large reserves of natural gas and coal still available. However, with the current consumption rate, the proven reserves of natural gas and coal should last for approximately 70 and 200 years, respectively, and oil is expected to deplete even earlier (Shafiee and Topal, 2009). By that time, an alternative fuel is needed for the future energy demand when those fossil fuels become unavailable, more importantly for the transport sector, which is consuming almost 60% of the world's energy (Balat and Balat, 2009).

Among potential candidates such as solar, wind, nuclear, tidal, hydro, biofuels, and geothermal energy, hydrogen appears to be the best choice due to the highest energy density per unit mass (120 MJ/kg), no environmental implications and its abundance in the universe (Dresselhaus and Thomas, 2001; Dunn, 2002; Tour et al., 2010). The need to balance the mismatch between energy supplied to the grid and the energy actually used from the grid by storing the excess energy is equally important to achieving a low carbon economy. It is against this backdrop that energy storage is believed to be essential in the modern energy supply chain as it will help to plug the leakages and improve efficiency. As a result of this, energy storage has recently attracted the attention of governments, stakeholders, researchers, and investors as it may be used to improve the performance of the energy supply chain (Box 12.2).

12.5.1 Energy storage (ES)

Energy storage is the capture of energy produced at one time for use at a later time. It involves converting energy from forms that are difficult to store to more conveniently or economically storable forms. A device that stores energy is generally called an accumulator or battery. Energy comes in multiple forms, including radiation, chemical, gravitational potential, electrical potential, electricity, elevated temperature, latent heat, and kinetic. Some technologies provide short-term energy storage, while others can endure for much longer. Bulk energy storage is currently dominated by hydroelectric dams, both conventional, as well as pumped.

Literally, energy storage occurs in every facet of human society. The fundamental process of photosynthesis through which green plants generate food involves the conversion of solar energy from sunlight to chemical energy, which is stored in plant cells. Storing fuelwood to provide heat during the winter or using it to maintain a fire is also a form of energy storage. Energy can also be stored as a commodity or used to process materials, which are storable. For example, energy can be used to purify dirty water, which can be stored as drinking water.

Energy comes in various forms, although it can be broadly classified into e primary and secondary forms of energy. Primary energy is regarded as those energy sources that only involve extraction or capture, with or without separation from contiguous material, cleaning, or grading, before the energy contained in it can be converted into heat or mechanical work (Øvergaard, 2008). They are usually found in nature. They include all energy forms that have not been subjected to any conversion or transformation process. Typical examples are crude oil, coal, biomass, wind, solar, tidal, natural uranium, geothermal, falling and flowing water, and natural gas.

On the other hand, secondary forms of energy include all energy forms, which occur as a result of the transformation of primary energy using energy conversion processes. Secondary energy forms are more convenient forms of energy as they can directly be used by humankind. They are also known as Energy Carriers (EC). Examples of secondary energy forms are electricity, gasoline, diesel, ethanol, butanol, hydrogen, and heat. In engineering terms, energy storage is focused on the concept of storing energy in the form in which it will be reused to generate energy whenever needed. It is required for a wide range of different times and size scales. The range of storage can be from capacitors that store as little of 1 W of energy for a few seconds to chemical compounds that can be used for grid-scale storage of several TW h of energy for years.

Generally speaking, primary energy serves to supply one of the three consumption sectors, transport, heat, and electricity. Energy storage must meet completely different requirements for each of these consumption sectors, and the different storage concepts and technologies have to integrate in a concerted manner to provide the basis of an energy system. The general concept behind secondary energy storage is to capture energy produced at one time for use later. The process of capturing the energy is generally regarded as the charging, while the process of releasing the energy to be used is regarded as the discharging. The energy is stored using different kinds of materials, which are commonly referred to as the energy carriers.

12.5.2   Benefits of energy storage

Energy storage is often called the "holy grail" of energy. It is regularly touted as a solution to fixing our aging power grid, a critical tool in increasing the spread of renewable energy, and a bridge between the needs of utilities and their customers. Energy storage has lots of benefits. It is important in energy management. It helps to reduce energy wastage and increase energy utilization efficiency (Chan et al., 2012; Abedin and Rosen, 2011) of process systems. Storage of secondary energy forms, such as heat and electricity, helps to reduce the quantity of primary energy forms (fossil fuels) consumed to generate them. This, in turn, not only lower CO_2 and other greenhouse gas emissions together with the associated global warming (Mahlia et al., 2014) but also help to conserve fossil fuels, which are believed to be exhaustible. It can also play a crucial role in increasing the penetration of renewable, clean and intermittent energy resources such as wind energy, solar energy, and marine tidal current to the grid (Zhou et al., 2012; Pardo et al., 2014;

Plebmann et al., 2013; Castillo and Gayme, 2014), as well as help in load shifting (Kousksou et al., 2013). Energy storage helps in power system planning, operation, and frequency regulation (Tan et al., 2012; Chen et al, 2009). It helps to maintain energy systems stability, improve power quality in micro-grid systems, as well as match demand with supply (Tan et al., 2012; Castillo and Gayme, 2014; Ibrahim et al., 2007).

Depending on factors such as a facility's location, utility rates, and electrical load, energy storage can be an ideal solution for facilities to cut energy bills. The cost of energy storage systems is constantly dropping, while the number of installed customer-sited energy storage systems is increasing rapidly. Regarding the environmental benefits, energy storage has many environmental benefits that can make it a valuable tool for meeting sustainability goals. By improving the overall efficiency of the power grid, storage accelerates the broader adoption of renewable energy. On a more local level, an energy storage system has no emissions, so it can be placed anywhere in a facility with no immediate environmental or air quality impacts.

12.5.3 Energy storage technologies

There are many technologies used for energy storage purposes. These technologies can be broadly classified according to the purpose for which the energy is stored. They include electrical energy storage and thermal energy storage. Electrical energy is regarded as one of the most readily available forms of energy. It is a common consumer good (Ibrahim et al., 2007) and ranked only second to oil in consumption in 2012 (IEA, 2014). Presently, the production of electricity is highly centralized with power plants located far from the end users. Grid load leveling is usually based on the prediction of daily and seasonal usage using historical trends. When production is not sufficient, peaking power plants such as gas turbines and hydroelectric systems are usually deployed to meet the shortfall. Due to the aforementioned factors, the storage of electrical energy has become a necessity. Electricity in its form is not storable. The only way through which it can be stored is by converting it into a more stable energy form, which is storable with the intent of transforming it back to electricity when needed. There are various technologies, which can be used to convert electricity to other forms of energy which can easily be stored. These technologies are regarded as electrical energy storage technologies and can be grouped as follows:

- Mechanical Energy Storage (MES)
- Chemical Energy Storage (CES)
- Electrochemical Energy Storage (EES)
- Superconducting Magnetic Energy Storage (SMES)
- Cryogenic Energy Storage (CrES)

Thermal Energy Storage (TES) is one of the most practiced forms of energy storage (Rismanchi et al., 2012; Yau and Rismanchi, 2012). TES systems consist of devices that are used to store electricity or other waste heat resources in the form of thermal energy, pending the time when they are used to meet energy needs. There are three thermal energy storage methods (Garg et al., 1985; Lane, 1983):

- Sensible heat storage
- Latent heat storage
- Thermochemical heat storage

Any given energy storage technology has some unique features or characteristics, which make it suitable for a particular energy storage application. These unique features help in the determination of the best energy storage technology to be adopted in any given circumstance.

12.5.4   Energy policies dependent upon EES technologies

With regards to Electrochemical storage, the electricity generation necessitates large-scale energy storage applications. To this end, the determining parameters of such EES technologies include cost, lifetime, efficiency, power, and energy density. Among the representative EES of electrochemical storage are the batteries. Batteries are considered more efficient compared to other EES in terms of scalability, efficiency, lifetime, discharge time, weight, and mobility of the system. Among batteries devices, the rechargeable lithium ion batteries (LiBs) are the most successful electricity storage devices, but their applicability is limited to small electronic equipment (Cho et al., 2015). Electrochemical energy storage systems sustain a wide spectrum of energy densities, ranging from 10s of W h/kg—for VRFB and Lead-acid—up to 1350 and 13,000 W h/kg—for Zh-air and Li-air—respectively (Cho et al., 2015).

Hydrogen-based energy storage systems (HESS) is proven one of the most promising energy storage techniques, since it can bridge major sectors of an energy system, such as transport and electricity. In parallel, HESS can reduce greenhouse gas emissions when coupled with a renewable energy source or low carbon energy technology. Moreover, HESS is effectively integrating large quantities of intermittent wind energy (González et al., 2015). The energy storage capacity of a HESS refers to the amount of electrical energy, which can be stored in the whole system. This energy storage capacity of a HESS is provided by the hydrogen storage system, thus hydrogen storage is a key-factor in the optimum design and operation of a HESS. Specifically, HESS from renewable energy necessitates suitable hydrogen storage systems that are commonly fueled by compressed gas and/or metal hydrides. Nevertheless, unlike other hydrogen storage applications, including vehicles, a lack of design and performance evaluation methodology for stationary applications for hydrogen storage, such as energy storage from RES, has been pointed out (González et al., 2015). Moreover, the comparison between the determining parameters of HESS and other EES technologies necessitates the calculation of the electrical energy available in the hydrogen storage system, in conjunction with a conversion device that transforms the content of chemical energy in hydrogen into electrical energy (González et al., 2015).

In the relevant literature that examined the in-field applicability of the HESS technology, the main criteria in selection, design, construction, and operation of a HESS technology could be applied to an experimental renewable energy storage plant and are based on hydrogen technologies and fuel cells. These general criteria of design and operation of a HESS technology can be outlined as follows (González et al., 2015):

- Safety criteria: safety in the construction and operation of the facility.
- Final use criteria: location and end-use of the facility.
- Operational criteria: operation and maintenance of the facility.
- Energy and exergy criteria: energy and exergy performance.
- Economic criteria: cost of investment in Operation and Maintenance (O&M), replacement, removal, and recycling, as well as the cost of decommissioning.

In the literature on the hydrogen fuel cells (Chen et al., 2009), hydrogen-based energy storage systems are receiving increasing attention at the present time, especially when cabled their integration with renewable power sources. The essential elements comprise an electrolyzer unit that converts the electrical energy input into hydrogen, the hydrogen storage system itself, and a hydrogen energy conversion system to convert the stored chemical energy in the hydrogen back to electrical energy. Furthermore, there is a range of approaches in producing hydrogen directly from thermochemical or photochemical processes. These technologies mainly use concentrated solar energy, although they are also earlier stages of development, such as solar-hydrogen, solar metal, and solar chemical heat pipe (Chen et al., 2009). Conclusively, there is a strong technical preference for the generator element to be a fuel cell system in order to capitalize on its high energy conversion efficiency. Nevertheless, this does not preclude the use of hydrogen burning internal combustion engines (Chen et al., 2009).

12.6 Conclusion

Electricity is one of the key resources of the 21st Century. Almost nothing can be manufactured, and no comforts of the modern world can be enjoyed without electricity. With dwindling fossil fuel reserves and the stark dangers of global warming, the world's energy focus is turning toward renewables to secure the energy of the future. Renewable energy is being globally addressed on two fronts. On the one hand, fundamental sciences are developing economically viable means to produce energy with less environmental damages. These energy sources (solar, wind, wave, and biomass) have remarkably different dynamic characteristics than traditional energy sources. Consequently, energy delivery from renewable sources presents a whole new set of challenges for both energy providers and users. The current generation is highly dependent on the nonrenewable sources of energy, which are fast exhausting and polluting. Thus, the need for reduction of the dependence on nonrenewable sources of energy arises. The use of alternative sources of energy reduces pollution and environmental hazards, as well as strengthens the power system.

References

Abedel-Raouf, N., Al-Homaidan, A.,A., Ibraheem, I.,B.,M., 2012. Microalgae and wastewater treatment. Saudi J. Biol. Sci. 257–275.

Abedin, A.H., Rosen, M.A., 2011. Closed and open thermochemical energy storage: energy and exergy-based comparisons. Energy 41, 83–92.

Alameh, K., Vasiliev, M., Alghmedi, R., Alam, M., 2014. Solar Energy Harvesting Clear Glass for Building-Integrated Photovoltaics, pp. 210–213.

Baisden, P.,A., Choppin, G.,R., 2006. Nuclear waste management and the nuclear fuel cycle. Encyclopaedia of Life Support Systems Radiochem. & Nuclearchem.

Balat, M., Balat, H., 2009. Resent trends in global production and utilization of bio-ethanol fuel. Appl. Energy 86, 2273–2282.

Bolonkin, A., 2004. Utilization of wind energy at high altitude. In: International Energy Conversion Engineering Conference at Providence, USA.

Byrne, J., Toly, N., Glover, L., 2006. Transforming Power Energy, Environment, and Society in Conflict. Transaction Publishers, pp. vii–xii.

Cass, S., 2009. Solar Power will make a difference—eventually. Sustain. Energy. Available at: https://www.technologyreview.com/s/414792/solar-power-will-make-a-difference-eventually/ (Accessed 27 July 2018).

Castillo, A., Gayme, D.F., 2014. Grid-scale energy storage applications in renewable energy integration: a survey. Energy Convers. Manag. 87, 885–894.

Chan, C.W., Ling-Chin, J., Roskilly, A.P., 2012. A review of chemical heat pumps, thermodynamic cycles and thermal energy storage technologies for low grade heat utilisation. Appl. Therm. Eng. 50, 1257–1273.

Chen, H., Cong, T.N., Yang, W., Tan, C., Li, Y., Ding, Y., 2009. Progress in electrical energy storage system: a critical review. Prog. Nat. Sci. 19 (3), 291–312.

Cho, J., Jeong, S., Kim, Y., 2015. Commercial and research battery technologies for electrical energy storage applications. Prog. Energy Combust. Sci. 48, 84–101.

Chow, T., Li, C., Lin, Z., 2010. Innovative solar windows for cooling-demand climate. Sol. Energy Mater. Sol. Cells 212–220. Elsevier.

Conn, R.,W., 2018. Nuclear Fusion. Available at: https://www.britannica.com/science/nuclear-fusion (Accessed 27 July 2018).

Divya, S., Jibin, J., 2014. A brief review on recent trends in alternative sources of energy. World Academy of Science, Engineering and Technology Int. J. Electr. Comput. Eng. 2 (8), 389–395.

Dresselhaus, M.S., Thomas, I.L., 2001. Alternative energy technologies. Nature 414, 332–337.

Dunn, S., 2002. Hydrogen futures: toward a sustainable energy system. Int. J. Hydrogen Energy 27, 235–264.

Ellis, Z., 2017. Industrial Waste Management. Library Press, USA.

Galvez, J.,B., Rodriguez, S.,M., Delyannis, E., Belessiotis, V.,G., Bhattacharya, S.,C., Kumar, S., 2010. Solar Energy Conversion and Photoenergy Systems: Thermal Systems and Desalination. Plants' EOLSS publications.

Garg, H.P., Mullick, S.C., Bhargava, A.K., 1985. Solar Thermal Energy Storage. D. Reidel Publishing Company, Dordrecht (Holland).

González, E.L., Llerena, F.I., Pérez, M.S., Iglesias, F.R., Macho, J.G., 2015. Energy evaluation of a solar hydrogen storage facility: comparison with other electrical energy sto- rage technologies. Int. J. Hydrogen Energy 40 (15), 5518–5525.

Greaves, D., Iglesias, G., 2018. Wave and Tidal Energy. John Wiley & Sons, p. 3.

Hadjipaschalis, I., Poullikkas, A., Efthimiou, V., 2009. Overview of current and future energy storage technologies for electric power applications. Renew. Sustain. Energy Rev. 1513−1522.

Howells, M., Alfstad, T., Victor, D., Goldstein, G., Remme, U., 2005. A model of household energy services in a low-income rural African village. Energy Policy 33 (14), 1833−1851.

Ibrahim, H., Ilinca, A., Perron, J., 2007. Energy storage systems − characteristics and comparisons. Renew. Sustain. Energy Rev. 12, 1221−1250.

IEA, 2014. Key World Energy Statistics. International Energy Agency.

IRENA, 2018. Renewable Power Generation Costs in 2017. International Renewable Energy Agency, Abu Dhabi.

Kalogirou, S.A., 2005. Seawater desalination using renewable energy sources. Energy Combust. Sci. 31 (2005), 242−281.

Kousksou, T., Bruel, P., Jamil, A., El Rhafiki, T., Zeraouli, Y., 2013. Energy storage: application and challenges. Sol. Energy Mater. Sol. Cells 120, 59−80.

Lane, G.A., 1983. Solar Heat Storage: Latent Heat Materials. CRC Press, Boca Raton (USA).

Li, M., 2014. World Energy 2014−2050. http://peakoilbarrel.com/world-energy-2014-2050-part-3 (Accessed 11 June 2019).

Liserre, M., Sauter, T., Hung, J.,Y., 2010. Future Energy Systems'. IEEE Industrial electronics Magazine.

Maehlum, M.,A., 2013. Tidal Energy Pros and Cons. Available at: http://energyinformative.org/tidal-energy-pros-and-cons/ (Accessed 27 July 2018).

Mahlia, T.M.I., Saktisahdan, T.J., Jannifar, A., Hasan, M.H., Matseelar, H.S.C., 2014. A review of available methods and developments on energy storage; technology update. Renew. Sustain. Energy Rev. 33, 532−545.

McClelland, J., 2015. Power from the People. Available at: https://www.eniday.com/en/human_en/power-from-the-people/ (Accessed 27 July 2018).

Øvergaard, S., 2008. Issue paper: definition of primary and secondary energy. In: Standard International Energy Classification (SIEC) in the International Recommendation on Energy Statistics (IRES). Oslo Group on Energy Statistics, pp. 1−7.

Panwar, N.,L., Kaushik, S.,C., Kothari, S., 2011. Role of renewable energy sources in environmental protection: A review. Renew. Sustain. Energy Rev. 1513−1524.

Pardo, P., Deydier, A., Anxionnaz-Minvielle, Z., Rouge, S., Cabassud, M., Cognet, P., 2014. A review on high temperature thermochemical heat energy storage. Renew. Sustain. Energy Rev. 32, 591−610.

Pereira, M., Freitas, M., da Silva, N., 2010. Rural electrification and energy poverty empirical evidences from Brazil. Renew. Sustain. Energy Rev. 14, 1229−1240.

Pittman, J.,K., Dean, A.,P., Osundeko, O., 2011. The Potential of Sustainable Algal Biofuel Production Using Wastewater Resources. Bioresource Technology, pp. 17−25.

Plebmann, G., Erdmann, M., Hlusiak, M., Breyer, C., 2013. Global energy storage demand for a 100% renewable electricity supply. Energy Procedia 46, 22−31.

Popescu, M.,F., 2015. The economics and finance of energy security. In: 22nd International Economic Conference − IECS 2015 "Economic Prospects in the Context of Growing Global and Regional Interdependencies". Romania.

Rismanchi, B., Saidur, R., Boroumandjazi, G., Ahmed, S., 2012. Energy, exergy and environmental analysis of cold thermal energy storage (CTES) systems. Renew. Sustain. Energy Rev. 16, 5741−5746.

Seboldt, W., 2004. Space-and earth-based solar power for the growing energy needs of future generations. Acta Astronautica 389−399. Elsevier.

Shafiee, S., Topal, E., 2009. When will fossil fuel reserves be diminished? Energy Pol. 37, 181–189.

Tan, X., Li, Q., Wang, H., 2012. Advances and trends of energy storage technology in microgrid. Electr. Power Energy Syst. 44, 179–191.

The Renewable Energy Hub, 2018. A Guide to BlueGen Micro CHP Co-generation System. Retrieved 22 August 2019, from. https://www.renewableenergyhub.co.uk/main/microcombined-heat-and-power-micro-chp-information/the-bluegenfuel-cell-microchp-co-generation-system/.

Tour, J.M., Kittrell, C., Colvin, V.L., 2010. Green carbon as a bridge to renewable energy. Nat. Mater. 9, 871–874.

Vidyasagar, A., 2016. What are Algae? Available at: https://www.livescience.com/54979-what-are-algae.html (Accessed 26 July 2018).

Waters, S., Aggidis, G., 2016. Tidal range technologies and state of the art in review. Renew. Sustain. Energy Rev. 514–529.

Willcox, S., 2012. A rising tide: the implications of climate change inundation for human rights and state Sovereignty [online] Essex Human Rights Rev. 9 (1). Available at: http://projects.essex.ac.uk/ehrr/V9N1/Willcox.pdf (Accessed 27 July 2018).

Yau, Y.H., Rismanchi, B., 2012. A review on cool thermal energy storage technologies and operating strategies. Renew. Sustain. Energy Rev. 15, 787–797.

Zhou, Z., Wu, W., Chen, Q., Chen, S., 2008. Study on sustainable development of rural household energy in northern China. Renew. Sustain. Energy Rev. 12, 2227–2239.

Zhou, Z., Benbouzid, M., Charpentier, J.F., Scuiller, F., Tang, T., 2012. A review of energy storage technologies for marine current energy systems. Renew. Sustain. Energy Rev. 18, 390–400.

13 ⠿

Rural electrification: field visits

A few years ago, the author visited five Indian villages and a village in Mozambique, where off-grid systems are being used. The observations and discussions with the electricity providers, as well as with the local population during these visits, have clarified a number of questions and issues.

This chapter provides a brief report of the site visits. It also provides a few suggestions for cost reduction and operational time saving by modifying procedures and maintenance practices.

13.1 India

The visit to India covered two husk power systems in the West Champaran District, Bihar, and three locations near Delhi, where solar-PV based systems are being used to provide electricity for the local population. The visit to the Husk Power Systems and the solar PV systems in Maukhas, Naglamal, and Murlipur were also covered. The locations offered contrasting pictures in terms of standards of living: at the solar-PV locations, which were close to Delhi, the villagers already had a connection to the main grid but received unreliable and a very limited supply of electricity. Also, the villages had stone and brick-built structures, showing better living conditions. In contrast, at the rice-husk plant locations, the majority of houses were built from straw, bamboo, and mud, with living conditions being quite poor.

The electricity provided from both the solar-PV and rice husk stations are both for the purpose of providing light. No other usage is possible, either due to the limited supply capacity or technology option. In the case of Solar-PV, unless solar cookers are included in the service, the possibility is limited.

With regard to the rice husk plant, consumers cannot exceed the allowed load limit (due to the fuse system or load limit installed). More importantly, the power station simply cannot provide any additional power than what had been originally allocated to each household, again, due to limited capacity. The two off-grid power systems, which use different sources of energy, have provided an insight into the working practice related to the generation of electricity within the above five Indian villages. The following is a summary of what has been observed during each visit.

13.1.1 Solar-PV power system

The first visit to Maukha, Naglamal, and Murlipur villages has provided some details on how the system has established itself and how it has been organized on a daily basis. The system is being used as a charging station (using batteries) to charge a number of solar lanterns on a daily basis. As has been reported by an NGO from Radha Govind Engineering College, Meerut, these solar lanterns are able to provide light for more than 50% of the households in these villages. The overall cost of the system, as has been reported by one of the working staff at the site, was approximately Rs 250,000. Out of this, Rs 230,000 was paid by an Indian university, and the balance of around Rs 20,000, contributed by an entrepreneur, has also been paid. As such, the university owns the system, while the entrepreneur will get a financial return, in the form of Rs 1 (or Rs 2 in another village with solar PV-station) per solar lantern being hired. The following is a brief description related to the system structure.

In a chosen household, the family hosting the system—where an investment has been made by the entrepreneur who invested in the system—will be operating it on a daily basis, as well as collecting the fees. In addition, the system's host family (entrepreneur) will obtain their light free, as part of the agreement. The villagers will collect their light (solar lantern) in return for advance payment of Rs 3. The money collected will be divided between the entrepreneur, NGO, and the university. The NGO's share will be used for basic maintenance. During the visit, it was noted that there were a number of solar lanterns located in rows and being charged with electricity from the solar-PV system directly and/or from the storage batteries. In this way, the lanterns will be fully charged and ready for collection at the end of the day.

As the solar PV system was small, there was a limit on the number of lanterns that could be provided to the rest of the village. The volunteer (from the NGO), who visits to check the system from time to time, mentioned that the demand for lanterns is much higher than what is presently available in the villages. The possibility of additional panels, as well as storage batteries, had been discussed among the visiting team, the technicians, as well as with the entrepreneur. The possibility of upgrading the system is a good idea, and something needs to be done as soon as possible. However, the investment needed, regardless of whether in the form of a grant from the government or from the university, may take a considerable time before any financial assistance can be made available. The system structure, however, as a whole is working well for the local community who is presently benefiting from it, in that the demand for solar lanterns are high, which in itself is a clear indication that the overall approach in establishing this kind of enterprise is a success, despite some of the shortages mentioned.

The development and expansion of the system could enable all households in the village to have access to solar lanterns. Additional panels and batteries could be installed, not necessarily within the same household, where the system originally had been located. Obviously, this kind of expansion and development can only be possible if funding is available. The technical know-how and human resources are already available for these villages. However, issues related to funding appear to be a major concern. To reduce technical inefficiency of the system, better quality Solar-PV cells should be considered. Higher Solar-PV efficiency may also help to reduce the house visit required by the volunteer.

13.1.2   Solar-PV system observations and changes

Observations During each meeting with the staff and the local population, the language used for communication was Hindi. Although some members of the visiting team did translate some of it, there was no simultaneous translation facility. This was a problem for a non-Hindi speaker, as was the case with the author. Intense and long conversations among different people were taking place in all the villages visited.

Another observation is related to the dust accumulating on the panel, which in itself reduces the efficiency of the system, to say the least. This issue was pointed out to one of the workers with a question connected to the frequencies in cleaning the panel. The worker informed me that although they clean the panel, the dust does accumulate very quickly on the surface.

The fixed position of the solar panel does not provide maximum exposure to the sunlight. However, a tracking system is likely to be costlier than a fixed system. Finally, it appears that the batteries used to store electricity from the solar panel are quite inefficient and have a short life-cycle.

Changes The following changes can be made, if and when additional funding is available. The investment will provide a huge saving in the long term. Other changes may not require additional funding, as these changes will focus on time management and work organizations, in relation to each power system.

1. The use of higher capacity batteries with a longer life cycle (examples of which are already available on the market than those presently being used in the villages).
2. Two options are available to constantly expose the solar panel to the sunlight as an alternative to the present fixed location on the roof. The first option is to have the panel shaped like a cone or sphere. That means regardless of the movement of the sunlight relative to the location, the panel (or part of it) will be always exposed to it. The position of the panel will be fixed on the ground. The other option can be achieved by making the solar panel mobile, able to move around its axis to follow the sunlight. This can be achieved by simply installing a light sensor(s) on the edge of the panel. A small motor will allow the solar panel to move slowly toward the new position, where a higher concentration of light is available. Light sensors and basic software to guide the movement are not expensive to obtain; however, this may seem very costly to achieve in India as the technology, as I was told, may not be available on the market in this part of the world.
3. Wiping the dust from the solar panel should be done regularly and on a daily basis. A clean solar panel will obtain light much more efficiently, and consequently, generate higher input of electricity than the present state of the panels observed in these villages.

13.1.3 Rice husk power system

The author traveled to the West Champaran district in Bihar State for the purpose of visiting the first rice husk power system in Tamkuha village. The following day, a second visit was made to Madhubani village, which was located a few miles from the first village. According to the information published on the company website (Husk Power Systems, 2018), the rice husk power system started its operation in August 2007, using unwanted rice husks as the main source for generating electricity. Since that time, the company has expanded its power supply to a large number of villages, around more than 250 villages/hamlets, which, as a result, has helped greatly in the reduction of fossil fuel usage, such as kerosene and diesel. A cofounder of the company "Husk Power Systems Pvt Ltd" explained that both plants have a 40 kVA engine, and they consume 50 kg/h of rice husk. Each station operates for 6 h per day. The husk consumption per plant/day is around three quintals (one quintal = 100 kg or 112 lb). Two persons are being employed on a daily basis to keep the power station in operation, with one worker as the "plant operator" and the second worker as a "husk loader." There is also an electrician, who is constantly moving between the rice husk power systems and the villages, mainly to deal with customers if there are any complaints. Electricity charges are collected in person each day. Also, each household has to provide a nonrefundable deposit of Rs 100 to the company before any supply of electricity can be made available. By-products, such as the charcoal left from the burned rice husks, have provided another business opportunity for the company and to the villagers as well. This by-product has been used to produce incense sticks to be sold in the local market (Boxes 13.1–13.3).

Another important observation is that there were no large proper storage facilities for the rice husks. This is mainly due to the fact that the supply of rice husk arrives on a daily basis, and consequently, large storage facilities are not an important factor at present. However, when the question about a possible shortage of rice husk supply was put forward to the cofounder, the answer was that the present storage method is sufficient for the time being and that there will be other sources of biomass materials to be used, such as elephant grass and bamboo, should there be a shortage of rice husk supply.

13.2 Rice husk system observations & changes

Observation There were two issues that could be reexamined in order to reduce costs and save time. The first issue is the daily routine in dismantling some of the hardware for the purpose of removing deposited tar after 6 h of operation. This kind of manual work is

---

BOX 13.1 Rural electrification weakest factors

The weakest points in the Technical & Energy Resources Area are 1. Maintenance requirement, 2. Metering system 3. System efficiency.

BOX 13.2 Aspects of rural electrification projects

According to the World Bank Independent Evaluation Group (IEG), rural electrification projects do not receive priority allocation for funding in some of the developing countries, as these governments believe that there are other more urgent priorities such as drinking water, health issues and roads, which need to be addressed first. For this reason, rural electrification projects, such as those implemented in Asia, have high costs with fewer benefits than the objectives outlined initially. The financing issue, therefore, is a major challenge in implementing and speeding up the construction of rural electrification projects. As the demand for energy increases and the prices of fossil fuel continues to rise, as well as the need to meet the new regulations related to the global warming, new ideas and new technologies are regularly introduced into our daily lives. This kind of approach, trying to solve the energy crisis worldwide, may have helped, in one way or another, those neglected and forgotten demands for energy in rural areas within many developing countries. Some of the tested approaches for the countryside is in the use of DG. Powering DG via conventional or nonconventional fuels for the purpose of generating electricity has been successfully achieved in many parts of the developing world by employing small-scale devices and local energy resources. The main purpose is to find a reliable and economical way of providing electricity needed for the local community.

Various types of DG technologies, their viability and their other important aspects, such as the economic and performance sides, are all vital factors within the overall success of any rural electrification project.

Positive and negative factors for each model should be examined whenever and wherever possible.

BOX 13.3 Hybrid system

1. Applying the strong points of multiple technologies
2. The hybrid system should be close to optimal in both performance and cost
3. The capital cost is high (e.g., cost of solar, cost of batteries, and cost of maintenance)
4. Better reliability of power supply
5. Need technical and engineering know-how
6. Ideal as a back-up power generating systems

time consuming, and consequently, may delay some of the work needed elsewhere, such as the development of the power systems in this location and/or in other areas, as well as the allocation of additional time in dealing with the end user needs. The second issue is the filter used during the cleaning process before the gas enters the 40 kVA engine. This can be changed to a more efficient filter, which can reduce the processing time and reduce/remove some of the mechanisms used to clean the gas. Also, the life cycle of the new filter can last a long period of time.

As regards the needs for a large permanent storage system, the co-founder indicated that the abundance of rice husk supply on a daily basis, as well as the availability of other types of energy crops within these villages, does not warrant a costly facility. A further reason is that the storage of rice-husk on a long-term basis may not be viable as the materials tend to degrade over a period of time. Storage facilities with controlled temperature can be costly and possibly beyond the present budget of the above power plant.

There was no metering system for each household; instead, if the household attempted to use electricity in excess of more than what is allowed/paid for, a fuse system will immediately come into operation and disconnect the power supply. According to the cofounder, a number of individuals tried to steal electricity by connecting their home directly to the mainline. This kind of theft, according to the cofounder, will not provide the additional power supply. This is because the system has been designed to provide electricity to a number of houses no higher than the system can cope with. Reportedly, some of the above illegal connections ended in injuries, as well as in one death.

A good practice noticed on the site was the case of regular training programs provided for the local population. These training lessons are vital as they prepare the local people to get involved with the project and may provide later on work for each participant. Finally, the comments received, if any, from the villagers were related mainly to the six hours of electricity supply, i.e., six hours of supply not being enough.

13.2.1   Changes

1. There are already various types of coating (nano-coating materials) available on the market, which can be used for the interior of the furnace. These types of coating can make the accumulation and the removal of the tar a much simpler and easier task than what had been witnessed.

2. There are already very efficient filters available on the market. These filters can provide higher purification to the gas, which, consequently, make the combustion engine more efficient and last longer. In addition, these newly produced gasification filters will reduce the number of stages used presently to clean the gas.

3. A large permanent storage for all the power systems located within a short distance from each other should be established as a reserve to cater needs in case of higher demands or sudden shortages of rice husk (or other types of biomass materials) supply.

4. A safer metering system could be installed rather than the present primitive method used in selling the electricity to each household, but the cost may be a concern here.

Finally, an important observation, which in the first instance may seem directly irrelevant to the electricity supply, is related to the conditions of the roads leading to these villages. Some of these roads are difficult to access and/or to drive a car on them, as in many cases, the road is simply made up of dust and stones, with large potholes. Some of these roads will endanger the life of the drivers and the rest of the car occupants, particularly if the driver is not experienced enough in driving through these kinds of roads. Investment, therefore, in order to improve access to a large number of Indian's villages should be the first priority to help, directly or indirectly in facilitating the supply of electricity to these villages.

13.3 Conclusion

For avoiding bureaucracies and delays, the quickest solution for the shortages of electricity in India's countryside is the direct involvement of the private sector, such as the case of the rice husk power station mentioned in this chapter. The success of the rice husk stations is mainly due to the cofounders in genuinely believing that such power stations are a vital business opportunity, which should succeed as the local population is in need of this kind of service. Also, it is a yardstick of personal success to those private individuals who invested part of their income in establishing these energy businesses. Having said that, government support in terms of electricity regulations, and as well as financial support for individual users and to those who supply it is still essential at this stage. The conclusion regarding this field visit, therefore, is that electricity supply, in particular in remote areas of the countryside, can be only sustained under the following conditions:

1. The involvement of the private sector
2. The availability and easy access all year round of energy sources
3. Regular training and awareness program for the local population as a way of finding employment for them, directly or indirectly with the newly established power stations
4. Power stations should use local resources, whenever this is possible, in running their business
5. Local power stations should have an open line to get constant updates on the latest development in the field of electricity generation technology, in general, and RE technology updates in particular. These updates will bring awareness of what is taking place to those currently dealing with the present power generation technology in remote areas of the countryside, thus cost reduction and time saving can be achieved
6. Solar-PV stations should be located in more than one household location. This is mainly to reduce shortages of light supply and to create competition within the village, which should, later on, lead on to better service and lower prices
7. Investment by the local or national government improves road access to the present villages with local electricity supply and to those villages where a proposal is presently being made to establish new power stations. The solar-PV and rice husk

projects are both successful in their approach in the way they have provided electricity to the five Indian villages. However, by simple observations and open discussions, a number of ideas and solutions presented itself for the purpose of improving the present service being provided to the local community. These observations and proposed changes discussed in this chapter should be followed in order to sustain and expand thefuture supply of electricity in an efficient way to all the inhabitants of these villages.

13.4 Mozambique

The author visited "Pungwe" village in Mozambique as part of the observations for an off-grid solar-PV system installed in the village. The system provides power and light for the local community, including for the provision of electricity to the village clinic, school, and for pumping water.

The discussions with the electricity providers, as well as the author's own observations, have provided an example of the state for the off-grid electricity supply in rural Mozambique.

As one of the poorest African countries, Mozambique's access to the electricity needs to be doubled in order to encourage economic development and improve the social and health program, especially within the vast areas of the countryside.

According to the Swedish International Development Cooperation Agency (Sida), Mozambique households are among the lowest in the region as regards connection to the main grid. This is because the percentage of the population who have access to electricity is 15% or less.

The main issue in regard to Mozambique's rural electrification stems from the low population density in the countryside, which discourages investments in these parts of the country, both from the private and the public sectors.

The visit to Pungwe village provided a good example of a rural electrification project in Mozambique, where villagers health, social, and economic prospects have improved drastically.

This section provides a brief outline of Mozambique's present and future electricity needs, as well as how the solar-PV project at Pungwe village can be improved to benefit further the local population. The possibility that an improvised version of the Pungwe electrification project can be copied at other locations of the countryside is another objective from this field visit.

13.4.1   Energy resources

Mozambique is rich with energy resources, such as hydropower, biomass, solar, wind power, natural gas, and coal.

The most suitable locations for generating electricity from hydropower in the country are located in areas such as Niassa and Manica. In regard to biomass, the areas around

Maputo and Sofala are two good examples. The most abundant solar energy locations are Tete, Nampula, and the south of Inhambane. During 2008 and 2012, NREL and The World Factbook reported that the potential of electricity generated from this source is around 2,477,570,614 MWh/year and the country's best locations for wind power are mostly located along the coastline, with Class 3–7 Wind at 50m (NREL, 2008). Cabo Delgado and Inhambane are the locations of the country's present natural gas resources. The official estimated reserve of the natural gas is 127,400,000,000 Cubic Meters (cu m) (The World FactBook, 2012).

Mozambique also has high coal reserves that have been estimated in millions Short Tons (*1 short ton = 907.18,474 kg*).

13.4.2   Electricity

In spite of the fact that Mozambique generates a large volume of electricity compared to the needs of the total population, not surprisingly, it is one of the lowest consumers of electricity in the world.

The main electricity for the nation is generated by the Cahora Bassa hydropower station, which generates around 2075 MW.

According to the "The World FactBook," the energy sources for the purpose of generating electricity have been estimated to be 2.9% sourced from fossil fuel and 97.1% from hydropower.

In spite of the fact that there are additional main power stations smaller than Cahora Bassa, the electricity supply to the countryside is still lagging behind. The irony is that a significant portion of the electricity generated by Cahora Bass hydropower is exported to South Africa. Some of the exported electricity from Cahora Bassa is reimported in order to supply electricity to the southern parts of the country, such as Maputo. Electricity transportation in this way is uneconomical as a significant percentage of electricity is lost during these transactions.

The above illustrates that the electricity generated in Mozambique is more than the country's requirements. The problem, therefore, is not the shortage of electricity, but rather the transmission and distribution of electricity.

Most of the electricity supplied is sourced from hydro and coal. During 2007, the percentage of electricity supplied from renewable energy sources was estimated to be less than 2% (Ahlborg and Hammar, 2011); however, the present situation is not that much different than 2007.

In regard to the electricity providers, there are two national entities where the electricity supply is concerned. The first is EdM (Electricidade de Mozambique), which supply electricity via the main grid to various parts of the country (EdM also controls the water supply system). The second entity is Funae (Fundo de Energia), mainly responsible for off-grid electrification.

According to USAID, the present domestic peak demand for electricity in Mozambique is more than 300 MW; however, similar to the domestic customers, the industrial customers are regularly subjected to frequent short outages and voltage fluctuations (USAID, 2008, 2019).

13.5 Generation capacity

- Installed Capacity: 2827 MW
  - Hydroelectric: 2184 MW
  - Thermal: 643 MW
- Power Africa New MW to Date Reached Financial Close: 280 MW
- Power Africa 2030 pipeline: 1947 MW

13.6 Connections

- Current Access Rate: 29%
  - Rural: 15% Urban: 57%
- Households without Power: 4.1 million
- Target: Universal Access by 2030
- Power Africa New Off-Grid Connections: 911 (USAID, 2019)

In regard to the private investments for the generation of electricity, this issue is one of the main challenges connected directly and indirectly to the progress of rural electrification in Mozambique. In spite of government encouragement for the private sector to get involved in supplying electricity, most of the present generated electricity is from governmental utilities.

13.6.1   Pungwe village

The village is located 150 KM from Maputo, the capital of Mozambique. After leaving the main highway, the access to the village is through an unpaved road for more than a 2-h drive. The standards of living within the village are relatively poor; however, by providing basic health service, schooling, and drinking water, it is higher than in other parts of rural Mozambique, according to Pungwe project leader.

All the power utilities installed within the village are solar-PV systems. These systems supply electricity for a number of buildings and locations within Pungwe.

Prior to the implementation of the present solar-PV project, villagers use petrol lanterns and candles to light their homes. The present generated electricity is used for the purpose of providing light, refrigeration, and pumping water.

The Pungwe project's overall approach has benefited the villagers in various ways, in spite of the negative comments made by some of the officials from the village committee as to the lack of money for the project and not enough technical assistants being provided by the government.

13.6.2   Pungwe electricity supply

As has been mentioned in the previous section, there are a number of solar-PV systems located in different parts of the village. Each system provides the electricity needed for a particular building or location, such as school, clinic, village café, water supply, and lighting. These systems, according to the project leader, have been commissioned by Funae.

There were no other types of renewable energy systems in operation in Pungwe village.

The proposal mentioned during the author visit is that all systems should be interconnected together in order to reroute electricity when it is not in use, either to be used in a different part of the village where it is needed and/or to charge some of the storage batteries with lower levels of energy than the rest of the batteries.

Neither an additional supply of electricity nor an emergency power supply is possible. This is because the consumers cannot exceed the electricity provided due to load limit, and more importantly, due to the absence of a proper distribution network board within the structure of the systems.

13.6.3   The clinic

The electricity obtained from the clinic solar-PV is used for operating the clinic drug fridge and for the provision of light. Consequently, the clinic can only deal with very basic health issues, such as minor individual injuries and vaccinations. There was one battery storage system on the side of the clinic compared to two storage battery systems observed in the other locations of the village (Fig. 13.1).

FIGURE 13.1 Pungwe clinic.

13.6.4   The school

Many schools in the countryside of Mozambique do not have access to electricity.

The selection process of which school should have access to electricity depended on whether or not the school would hold classes in the evening. These evening classes are held due to the limited classroom space in relation to the number of students attending the school.

Therefore, in order to install a solar-PV system in a rural school, this may depend on the school's ability and resources before actual consideration can be made by the local authority.

In regard to Pungwe village, all the classrooms' lighting is provided via the solar-PV installed close to the main building. Each classroom has been provided with two bulbs. There were two storage batteries installed at the entrance of the school (Figs. 13.2 and 13.3).

13.6.5   Water pumping

According to the village committee, prior to the installation of the solar-PV systems in the village, there was no clean drinking water in Pungwe, as there was no electricity supply for this purpose.

There is one well within the village where clean drinking water can be obtained. At present, drinking water is available to all the village populations, due to the installed pumping filtered devices operated via the electricity generated from the solar-PV system (Fig. 13.4).

FIGURE 13.2 Classroom Pungwe school with two solar-PV lighting bulbs.

FIGURE 13.3 Two storage batteries for each main location within the village of Pungwe.

FIGURE 13.4 Water storage for drinking water pumped from the village well and filtered using electricity generated via the solar-PV system.

13.6.6   Systems data

Regarding the amount of power generated via all the village solar-PV systems, the author did not manage to get the data from each solar-PV system, nor the overall number of MW generated from all the systems combined together, despite the repeated requests made during the visit. Various technical details, as the author informed, were not available at that particular moment. However, observations related to the systems' electricity output for domestic use can be estimated to be around 100 Wh (or less) per household. The overall total can be estimated by knowing the exact number of households within the village.

13.6.7   Solar-PV system observations and changes

Observations The language used with the staff and the local population was mainly English; however, Portuguese was also used on many occasions, especially when talking to the local residents or with some of the staff who were not fully familiar with the English language.

Another observation is related to the solar panels, most of which were relatively clean, despite the dusty/sandy nature of the village soil. This kind of dust-free panels should increase the efficiency of the system and reduce the possibility of damages occurring to them.

All the solar panels have been designed in fixed positions. This kind of design does not provide maximum exposure to the sunlight. A tracking system is the best option where panels can be exposed to the sunlight throughout the day. In regard to the storage batteries, the project leader who was providing various details about the village and the solar-PV systems could not provide technical details; however, under close inspection of the batteries, all looked in good condition and possibly recently installed.

Changes The author was informed that the solar-PV for Pungwe village is fully subsided by the government; however, according to the project leader, the money allocated to the project was not sufficient for the overall cost.

The project leader mentioned that he visits the village every 3 months to inspect the systems and sort out any problem related to the electricity supply within the village.

There were a number of questions and suggestions from the author in regard to the solar-PV systems. Most of the questions and suggestions were discussed and debated with the project leader and other members of the village committee.

In regard to general observations and suggested changes to the systems, there are a number of points, which can benefit the present project. Some of these points are similar to the suggestions made during the visit to India's villages, where solar-PVs systems are in use. The exception here is that the private investment for rural electrification in Mozambique is not noticeable, if not completely absent, from the location the team visited, contrasting with the case in India where local enterprisers are involved within the investment schemes of the villages. Pungwe village, therefore, is one example, which illustrates the complete lack of private investments (from the local population and outside it).

Corruption at various levels in the development of various projects and businesses (regardless of whether it is related to the generation of electricity or not) is another major obstacle to the development of the countryside in general and the development of the electricity sector in particular. It has been estimated that around 5% of the countryside in Mozambique has access to electricity; therefore, the following points should be considered as it can be applied to Mozambique, in general, and Pungwe village in particular:

1. Higher capacity batteries with a longer life cycle should be used within the present locations, as well as in other locations of the village where power supplies are still missing
2. The solar-PV systems should be connected together with an up-to-date controlling board mechanism to distribute electricity whenever needed or to store it at locations where mostly needed
3. All solar panels should be mobile (or shaped like a cone or sphere), i.e., available to constantly to receive the sunlight during the day
4. Private investments and further involvement of the local population should be the main aim for the Mozambique rural electrification projects in the countryside;
5. Additional renewable energy systems, such as wind turbine, biomass energy, and hybrid combinations of these systems should be implemented in order to sustain and provide a constant supply of electricity to the local population
6. As the cost of electrifying remote areas of the countryside is the main issue, SWER (single-wire earth return) can be used as a means to reduce the overall cost and to connect to the main grid, whenever this is possible
7. An awareness program for the local population to help in establishing energy cooperative enterprises within the villages where it is not possible to connect to the main grid
8. Investment by the local or national government to improve road access to the present villages with local electricity supply and to those villages where a proposal is presently being made to establish new power stations
9. Microhydro systems and solar-PV systems can be the economic approach when the cost is the main issues when it comes to off-grid systems for remote rural areas of Mozambique
10. Regular systems maintenance should be made as part of an enforced timetable rather than whenever the breakdown takes place.

13.6.8   Conclusion

The country has abundant sources of natural energy in the countryside, which can be exploited for the benefit of all the local population, providing access to electricity without a negative effect on the environment. However, the overall picture obtained from this visit is that, in principle, there are no shortages of electricity, but rather there are "shortages" of access to electricity in Mozambique. Exporting electricity, then importing what is needed in the form of electricity, surely is not the ideal solution for Mozambique.

When it comes to solar-PV electrification projects, the positive outlook for Mozambique is within the active role the National Energy Fund in supporting solar panels for rural electrifications. However, there are many shortages related to the electricity transmission and distribution, which need to be dealt with, particularly where governmental law, regulations, and practical implementations are concerned. Financial support for individual users, as well as to those who supply it, is still essential at this stage. Electricity supply in remote areas of the countryside can be sustained in the form of:

1. Incentives to encourage the involvement of the private sector, in particular the local population, to get involved in financing and taking part in the rural electrification projects
2. Training and awareness program for the local population;
3. SWER should be one of the options as a way of reducing the overall cost of rural electrification
4. Renewable energy applications should not be limited to one type of system. In addition to the above, hybrid systems should be the main focus whenever this is environmentally and economically applicable;
5. Improving road access to towns and villages
6. Increasing the economic productivity via the direct usage of electricity already generated rather than exporting the above, particularly that the price of electricity unit generated is relatively low compared to other parts of the developing world
7. Off-grid projects with renewable energy sources should be the main aim and objective for the local and national institutes for rural electrification rather than the usage of diesel engines and/or other forms of generating electricity from fossil fuels
8. Competition should be encouraged by dividing big public electricity producers into smaller companies in order to create genuine competition for the benefit of the local population and in order to raise the quality of the service being provided
9. Maintenance issues and systems upgrade should be part of the regular working schedules of the project

References

Ahlborg, H., Hammar, L., 2011. Drivers and Barriers to Rural Electrification in Tanzania and Mozambique — Grid Extension, Off-Grid and Renewable Energy Sources. Policy Issues (PI), World Renewable Energy Congress 18-13 May, 2011. Linkoping, Sweden.

Husk Power Systems Pvt Ltd, 2018. Rice Husks. http://www.huskpowersystems.com/index.php (Accessed 11.11.2018).

NREL, 2008. Solar Resources by Class Per Country. Cited at Open EI. http://en.openei.org/datasets/node/498 (Accessed 4.5.2014).

Swedish International Development Cooperation Agency, n.d. Rural Electrification in Mozambique.

The World FactBook, 2012. Central Intelligence Agency (CIA), USA https://www.cia.gov/library/publications/the-world-factbook/fields/2045.html (Accessed 4.7.2014).

USAID, 2008. Presentation to CTA: Electricity Pricing Assessment for Mozambique Presented by Donald Hertzmark - Nathan Associates. http://www.speed-program.com/library/resources/tipmoz_media/cat3_link_1216528441.pdf (Accessed 4.3.2015).

USAID, 2019. Mozambique - Power AFRICA Fact SHEET. https://www.usaid.gov/powerafrica/mozambique (Accessed 04.03.2020).

Author index

Note: 'Page numbers followed by "f" indicate figures, "t" indicate tables and "b" indicate boxes.'

Subject index

Note: 'Page numbers followed by "f" indicate figures, "t" indicate tables and "b" indicate boxes.'

A

Aframax, 60
African Development Bank (AfDB) group,
     177–178
African Development Fund, 177–178
Air-conditioning system, 5
Algal biofuel, 237
American Petroleum Institute (API) gravity,
     48, 49b, 50
Anglo American plc, 24
Anthracite, 20b–21b, 29–30
Anticline petroleum, 39, 40f
Arch Coal, 24
Asian Development Bank (ADB), 149–150,
     158–159, 176–177, 176b–178b
Aviation industry, 75
Aviation jet fuel, 74, 75b

B

Backwardation market, 53, 54f
BHP Billiton, 23
Big Oil, 106
Biofuels, 100–101
     algal, 237
Biomass conversion systems, 128, 129b
Biomass energy, 4, 125, 127–129, 128f, 129b
Bituminous coal, 20b–21b, 29–30
Black shale rock, 39
Brent Blend, 50
Brent price–price cycles, 215, 217f
Building-integrated photovoltaics
     (BIPV), 239
Build-Own-Operate-Transfer (BOOT), 98

C

CAGR. *See* Compound Annual Growth Rate
     (CAGR)

Capital expenditure, 51, 85
Capital investment, 180
Caprock formation, 39
Carbon dioxide (CO_2) emission
     coal, 35–36, 35f
     reduction, 3, 15–16
Carbon economy, 241
Carboniferous Period, 20b–21b
Carbon sequestration, 235
CCoW. *See* Coal Contract to Work (CCoW)
CGAP. *See* Consultative Group to Assist the
     Poor (CGAP)
Chemical energy storage (CES), 243
CIS. *See* Commonwealth of Independent
     States (CIS)
Civilizations, types, 1
Climate change, 3, 4b, 16, 34
     coal policies, 26
     geopolitics, 23
     impacts of, 10–12
     OECD countries, 31
Climate Change Act, 232–233
Coal
     cleaner-burning fossil fuel, 19–20
     climate change, 34
     coal-to-gas fuel, 34
     CO_2 emissions, 35–36, 35f
     exploration, 22, 22f
     fiscal systems, 24, 25b
     formation, 20–21, 20b–21b
     geography, 23
     geopolitics, 23–24
     global market
         characteristics, 29–30
         climate change policies, 26
         demand, 26
         economic growth, 25

Printed in the United States
By Bookmasters